ALGEBRA II
ESSENTIAL PRACTICE WORKBOOK

HIGH SCHOOL

AMERICAN MATH
ACADEMY

By H. TONG, M.Ed.

Math Instructor & Olympiad Coach

www.americanmathacademy.com

AMERICAN MATH ACADEMY

ALGEBRA II
ESSENTIAL PRACTICE WORKBOOK
HIGH SCHOOL

Writer: H.Tong
Copyright © 2021 The American Math Academy LLC.

Printed in United States of America.

ISBN: 9798869116581

Although the writer has made every effort to ensure the accuracy and completeness of information contained this book, the writer assumes no responsibility for errors, inaccuracies, omissions or any inconsistency herein. Any slighting of people, places, or organizations is unintentional.

Questions, suggestions or comments, please email:americanmathacademy@gmail.com

TABLE OF CONTENTS

About the Author

Mr. Tong teaches at various private and public schools in both New York and New Jersey. In conjunction with his teaching, Mr. Tong developed his own private tutoring company. His company developed a unique way of ensuring their students' success on the math section of the SAT. His students, over the years, have been able to apply the knowledge and skills they learned during their tutoring sessions in college and beyond. Mr. Tong's academic accolades make him the best candidate to teach SAT Math. He received his master's degree in Math Education. He has won several national and state championships in various math competitions and has taken his team to victory in the Olympiads. He has trained students for Math Counts, American Math Competition (AMC), Harvard MIT Math Tournament, Princeton Math Contest, and the National Math League, and many other events. His teaching style ensures his students success. He personally invests energy and time into his students and sees what struggling with. His dedication towards his students is evident through his student's achievements.

Acknowledgements

I would like to take the time to acknowledge the help and support of my beloved wife, my colleagues, and my students–their feedback on my book was invaluable. I would like to say an additional thank you to my dear friend Robert for his assistance in making this book complete. Without everyone's help, this book would not be the same. I dedicate this book to my precious daughter Vera and Nora who were my inspiration to take on this project.

SAMPLE TEST

1) $\dfrac{x+2}{5} = \dfrac{x-3}{4}$, Solve for x.

2) $-3x - 18 + 20 \leq 2(x - 8)$, Solve for x.

3) If $ab = b + c$, Solve for b.

4) $\left|\dfrac{1}{2}x\right| \leq 9$, Solve for x.

5) Simplify $(6x^2 - 2x) - (2x - 5x^2)$.

6) Simplify $\dfrac{25x^{-8}y^{10}}{5x^{10}y^{-5}}$.

7) Factor $x^3 + 4x^2 - 3x - 12$.

8) Simplify $\dfrac{x^2 - 25}{x^3 - 125}$.

9) Simplify $\dfrac{x+1}{x^2-1} - \dfrac{x-1}{x^2+1}$.

10) Simplify $\dfrac{y}{y-1} \div \dfrac{y^2}{y^3-y}$.

11) Simplify $\dfrac{x}{\dfrac{3}{x} - \dfrac{5}{x}}$

12) If $g(x) = 3x - 5$, find $g(-3x)$.

13) Solve $9^{2x-3}=81^{3x-8}$.

14) Convert $\dfrac{\pi}{5}$.

15) Simplify $\sqrt{\dfrac{64x^6y^8}{16xy}}$.

16) Simplify $\sqrt{48}-\sqrt{12}$.

17) Simplify $\dfrac{\sqrt{x}+\sqrt{y}}{\sqrt{x}-\sqrt{y}}$.

18) Simplify $(x^{-10}y^5)^{-\frac{1}{5}}$.

19) Simplify $i^2 - i^{12} + i^{18}$.

21) If one term of arithmetic sequence $a_{48} = 120$ and the common difference $d = 12$. Find the first term.

20) How many liters of 80% pure water must be added to 40 liters of 30% pure water to produce 60% pure water?

23) From the equation of circle below find the radius.

$x^2 + y^2 + 2x - 6y = 20$

22) Find the sum of the infinite geometric series $5 + 10 + 15 + 20 + 25 + ...$

25) If matrix $x=\begin{bmatrix} 1 & 3 \\ -2 & 4 \end{bmatrix}$ and matrix $y=\begin{bmatrix} 1 & 3 \\ -2 & 4 \end{bmatrix}$, then find $x \cdot y$?

24) What is the value of x in $\log_2 3x - 1 = 5$.

SOLVING LINEAR EQUATIONS

A linear equation in one variable can be written in the form of y=mx+b, where m, y and b are any numbers and m≠0. An equation is linear when the power of x is 1.

How to Solving Linear Equations in One Variable Step by Step:

- ✓ First use distributive property rule to remove any parentheses.
- ✓ Add or subtract terms to get the variable on one side.
- ✓ Multiply or divide to solve for the variable.
- ✓ Check your answers.

Example:

$5(2x + 3) - 1 = 3(x + 15) - 45$

Solution:

$5(2x + 3) - 1 = 3(x + 15) - 45$	Apply the distributive property
$10x + 15 - 1 = 3x + 45 - 45$	Combine like terms.
$10x + 14 = 3x$	Place x–terms on one side and simplify.
$7x = -14$	Divide both sides by 7
$x = \dfrac{-14}{7} = -2$	

Check:

$5(2x + 3) - 1 = 3(x + 15) - 45,$

$5(2 \times -2 + 3) - 1 = 3(-2 + 15) - 45$

$5(-4+3) - 1 = 3(13) - 45$

$5(-1) - 1 = 39 - 45$

$-6 = -6$

SOLVING LINEAR EQUATIONS EXERCISES

1) $5x - 15 = 25$

2) $3x - 30 = 45$

3) $x - 10 = -25$

4) $x + 30 = -15$

5) $-2x - 5 = -35$

6) $-3x + 3 = -18$

7) $\dfrac{x}{2} + 3 = -4$

8) $-\dfrac{x}{4} - 5 = -9$

9) $\dfrac{x}{5} - \dfrac{3}{4} = -1$

10) $-\dfrac{x}{2} - \dfrac{7}{3} = -11$

11) $5x - 2(x + 5) = 4(2x + 6) + 8$

12) $6x - 11 = 5x + 19$

13) $\dfrac{x+2}{5} = \dfrac{x-3}{4}$

14) $\dfrac{x+3}{2x} = \dfrac{2}{3}$

15) $-3(2x + 7) = 4(5x + 10)$

16) $\dfrac{1}{2}(2x+10) = -4(5x + 10)$

17) $\dfrac{3x-7}{5} - \dfrac{x}{3} = 2 - \dfrac{x-3}{4}$

18) $\dfrac{1}{x} = \dfrac{3x+10}{4x}$

SOLVING LINEAR INEQUALITIES

Inequality Symbols	Number Line Symbols
$\leq$	←———•
$\geq$	•———→
$<$	←———○
$>$	○———→

Study Tips:

1. If $a < x$ and $x < b$, then $a < x < b$

2. If $a \leq b$, then $a < b$ or $a = b$

3. If $a \geq b$, then $a > b$ or $a = b$

4. If $a < b$ and $c < d$, then $a + c < b + d$

5. If $\dfrac{1}{a} < \dfrac{1}{x} < \dfrac{1}{b}$, then $a > x > b$ (a, x and b are positive integers)

SOLVING LINEAR INEQUALITIES EXERCISES

Solve the following system of linear inequality for x.

1) x – 15 < 25

2) 3x – 6 < 30

3) 3(x – 2) ≥ 18

4) 4(2x – 5) ≥ 3(2x + 10)

5) –10 < 2x – 12 < 20

6) 8 ≤ 4x – 20 ≤ 16

7) 10x – 15 (2x + 1) > 25

8) 12x – 3(x – 5) ≤ 42

9) $\dfrac{x+1}{3} \le \dfrac{x-3}{4}$

10) $\dfrac{x+4}{x} \ge \dfrac{2}{3}$

11) 7x –8 ≥ 3(x – 6)

12) 3x – (x + 5) > –2(2 + x) + 7

13) $\dfrac{1}{3}$(6x – 12) + 6 ≤2x –8

14) 3x – 5 + 4x ≥ – 4(x + 4)

15) x – 18 > –8

16) 5x – 10 < 30

17) –3x – 18 + 20 ≤ 2(x – 8)

18) –7x – 14≥ –14 – 2(x – 10)

SOLVING LITERAL EQUATIONS

Literal equation is an equation with two or more variables and can be rearranged to solve for a variable.

Example: The area of the trapezoid is $A = \dfrac{(b_1 + b_2) \cdot h}{2}$ what is the height in terms of the base and the area.

Solution:

If $A = \dfrac{(b_1 + b_2)h}{2}$ (cross–multiply)

$2A = (b_1 + b_2)h$ (divide each side by $b_1 + b_2$)

$\dfrac{2A}{b_1 + b_2} = h$

Example: Solve $a = \dfrac{bc}{rc + k}$ for c.

Solution:

$a(rc + k) = \dfrac{bc}{rc + k}(rc + k)$ multiply both sides of the equation by $(rc + k)$

$a(rc + k) = bc$ (distribute a)

$arc + ar = bc$ (subtract arc from both sides to get all c's on the same side)

$ -arc -arc$

$+\ \underline{}$

$ar = bc - arc$ factor out the c from each term

$ar = c(b - ar)$ divide both sides by $(b - ar)$ to isolate c

$\dfrac{ar}{b - ar} = c$

SOLVING LITERAL EQUATIONS EXERCISES

1) If $a(b + c) = d$, Solve for c.

2) If $\dfrac{1}{a} = \dfrac{1}{b} + \dfrac{1}{c}$, Solve for b.

3) If $4x - 3y = 6$, Solve for x.

4) If $V = \pi r^2 h$, Solve for r.

5) If $P = 2L + 2W$, Solve for W.

6) If $y = mx + b$, Solve for x.

7) If $x = 2(y - 3x)$, Solve for y.

8) If $xy + 3 = m$, Solve for y.

9) If $\dfrac{x+a}{3} = \dfrac{x-a}{4}$, Solve for a.

10) If $R = \dfrac{R_1 R_2}{R_1 + R_2}$, Solve for R_1.

11) If $\dfrac{x}{3} = \dfrac{y}{4}$, Solve for y.

12) If $A = \dfrac{1}{2} bh$, Solve for h.

13) If $E = mc^2$, Solve for c.

14) If $ab = b + c$, Solve for b.

15) If $c = \dfrac{5}{9}(f - 32)$, Solve for f.

16) If $x = \dfrac{1}{2}y - 2$, Solve for y.

ABSOLUTE VALUE EQUATIONS AND ABSOLUTE VALUE INEQUALITIES

Absolute Value: The distance of integers from zero on the number line.

- If $|x| = a$, then either $x = a$ or $x = -a$

- If $|ax + b| = k$, then either $ax + b = k$ or $ax + b = -k$

- If $|ax + b| < k$, then either $ax + b < k$ or $ax + b > -k$

- $|ax + b| \leq k$, then either $ax + b \leq k$ or $ax + b \geq -k$

Inequality Symbols

Symbol	Meaning	Number Line	Circle
$<$	Less than	←———o	Open circle
$>$	Greater than	o———→	Open circle
$\leq$	Less than or equal to (At least)	←———●	Close circle
$\geq$	Greater than or equal to (At most)	●———→	Close circle

Solve and simplify each of the following.

1) $|x - 5| = 12$

2) $|2x - 8| = 24$

3) $\left|\dfrac{x}{2} - 4\right| = 7$

4) $\left|\dfrac{2}{x} + 3\right| = 5$

5) $|x - 9| = |2x - 1|$

6) $\left|\dfrac{1}{2}x + 5\right| = 9$

7) $\dfrac{3^3 \times 3^{10}}{81^3}$

8) $\dfrac{2^4 \times 2^{12}}{32^3}$

9) $\left|\dfrac{3x}{2} - 5\right| = 12$

10) $|x + 5| = -9$

11) $|x - 9| \leq 12$

12) $\left|\dfrac{1}{3}x + 3\right| \leq 9$

13) $|3x - 5| \leq 5x$

14) $\left|\dfrac{1}{2}x\right| \leq 9$

15) $\left|\dfrac{3x}{2} + 5\right| - 8 = 18$

16) $\left|\dfrac{2x}{3} - 5\right| < 12$

PROPORTION, DIRECT VARIATION AND INVERSE VARIATION

Proportion: A proportion is an equation that shows two equivalent ratios.

If $\dfrac{a}{b} = \dfrac{c}{d}$, then, by cross multiplication, $ad = bc$.

Direct variation: A direct variation is a relationship between two quantities of x and y that can be written as the following form:

- $y = kx$, where k is a constant of veriation and k does not equal zero.

Inverse variation: an inverse variation is a relationship between two quantities of x and y that can be written as the following form:

- $xy = k$, where k is a constant of variation and K does not equal zero.

Example:
Find y when x = 3, if y varies directly as x and y = 20 when x = 5.

Solution:
If $y = kx$, then $20 = 5k$, $k = 4$
From $y = kx$, $y = 4x$, since $x = 3$ then, $y = 4 \times 3 = 12$

Example:
If y varies inversely as x and x = 12 when y = 60, find y when x = 18.

Solution:
If $y = \dfrac{k}{x}$ then $60 = \dfrac{k}{12}$, from cross–multiplication $\quad k = 60 \times 12 = 720$

From $y = \dfrac{k}{x}$, $y = \dfrac{720}{x}$, since $x = 18$ then, $y = \dfrac{720}{18} = 40$

Find the missing number in each of the following.

1) $\dfrac{x+4}{3} = \dfrac{x-5}{4}$

2) $\dfrac{x-8}{5} = \dfrac{x+4}{3}$

3) $\dfrac{\frac{x}{2}}{\frac{1}{5}} = \dfrac{x-6}{2}$

4) $\dfrac{x}{5} = \dfrac{2x+4}{8}$

5) $\dfrac{x+10}{2} = \dfrac{3x-5}{4}$

6) $\dfrac{2x-1}{3} = \dfrac{3x+1}{4}$

7) Find y when x = 3, if y varies directly as x and y = 20 when x = 5.

8) If y varies inversely as x and x = 12 when y = 60, find y when x = 18.

Identify each of the following as direct variation or inverse variation.

9) $y = \dfrac{7}{x}$

10) $y = 5x$

11)

x	y
1	4
2	8
3	12
4	16
5	20

12)

x	y
1	36
2	18
3	12
4	9
6	6

SOLVING SYSTEMS OF EQUATIONS BY SUBSTITUTION & ELIMINATION

Substitution Method:

Example: Solve the following system of equations using the **substitution method.**

$x - 2y = 18$

$2x + 3y = 15$

Solution:

Step 1: Solve for one variable of the equations (either y or x).

If $x - 2y = 18$ then $x = 2y + 18$

Step 2: Substitute this expression into the other equation.

$2x + 3y = 15$

$2(2y+18) + 3y = 15$

Step 3: Solve the equation for y.

$2(2y+18) + 3y = 15$

Then,

$4y + 36 + 3y = 15$

$7y = 15 - 36$

$7y = -21$

$y = -3$

Step 4: Substitute –3 for y in the equation.

$x = 2y + 18$

$x = 2(-3) + 18$

$x = -6 + 18$

$x = 12$

Solution (12, –3)

Elimination Method:

Example: Solve the following system of equations using the **elimination method.**

$3x - 2y = 24$

$2x + y = -5$

Step 1: Multiply the second equation by 2 and keep the first equation the same.

$2(2x + y) = -5(2)$ then...

$4x + 2y = -10$

Step 2: Add the revised equations and solve for x.

$3x - 2y = 24$

$4x + 2y = -10$

$+$ ——————————

$7x = 14$ then $x = 2$

Step 3: Substitute the value of x into one of the equations and then solve for y.

$3x - 2y = 24$ if $x = 2$ then

$3(2) - 2y = 24$

$6 - 2y = 24$

$-2y = 24 - 6$

$-2y = 18$

$y = -9$

Solution $(2, -9)$.

SOLVING SYSTEMS OF EQUATIONS
BY SUBSTITUTION & ELIMINATION EXERCISES

Solve each system by Elimination

1) $2x + y = 20$

$3x - 4y = 19$

2) $x + 2y = 15$

$3x - 2y = 17$

3) $3x + 2y = 10$

$x + y = 14$

4) $2x + 6y = 18$

$2x - y = 4$

5) $x - \frac{1}{2}y = 16$

$x + \frac{1}{2}y = 6$

6) $7x - y = 118$

$2x - y = 8$

Solve each system by Substitution

7) $x = 2y + 2$

$x - 4y = 8$

8) $x + 2y = 18$

$y = x + 6$

9) $x = 3y$

$x + y = 20$

10) $2x - 3y = 0$

$2x + y = 12$

11) $x - \frac{1}{2}y = 25$

$x = 3y$

12) $\frac{1}{2}x = 3y$

$x + 3y = 36$

Solve the following equations.

13) $3k - 5 + 4k = -4(k + 4)$

14) $\frac{5}{8}(8x + 24) = 4(x - 3)$

15) The perimeter of a rectangular garden is 48 cm. The width of the garden is 3 cm longer than 2 times the length. What is the length of the garden?

16) The cost of 4 child movie tickets and 2 adults movie tickets is \$48. If the price of each child's ticket is one–third the price of each adult's ticket, what is the cost for 1 child ticket?

Key Notes:

Parallel lines (No solution):

The equations have the same slope and different y–intercepts. The graphs are parallel.

$$\left.\begin{array}{l} d_1 : y = m_1 x + n_1 \\ d_2 : y = m_2 x + n_2 \end{array}\right\} \quad d_1 \mathbin{/\!/} d_2 \Longleftrightarrow m_1 = m_2$$

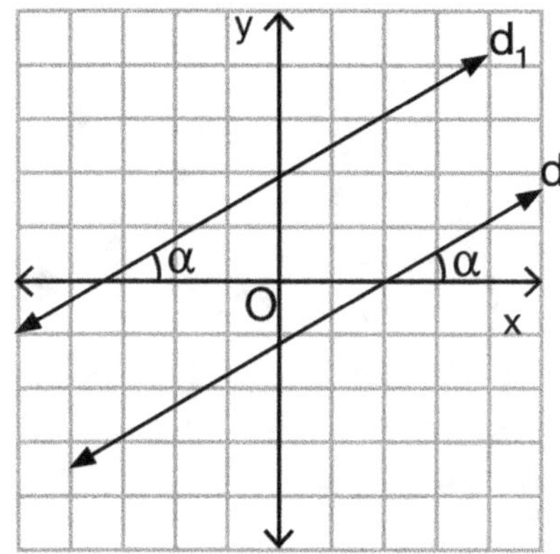

– Same slope $(m_1 = m_2)$

– Different y–intercepts

Example:

$2x + y = 3$
$6x + 3y = 12$

Perpendicular lines:

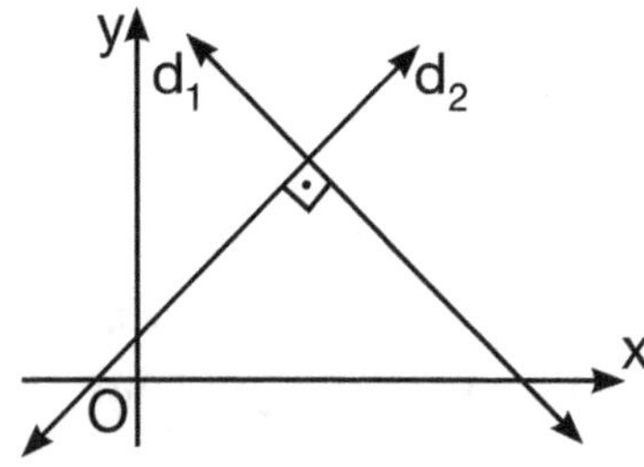

– Slopes are negative reciprocals of each other.

$$\left(d_1 \perp d_2, \text{then } m_1 = -\frac{1}{m_2} \right)$$

SOLVING EQUATIONS INVOLVING PARALLEL AND PERPENDICULAR LINES EXERCISES

Determinate whether the following pairs of lines are parallel, perpendicular, or neither.

1) $2x - 3y = 12$

$y = \dfrac{3x}{12} + 12$

2) $\dfrac{2x}{3} + y = 6$

$3x - 2y = 18$

3) $3x - 4y = 12$

$y = \dfrac{-4x}{3} + 12$

4) $x + y = 6$

$x = -y$

5) $\dfrac{1}{2}x + \dfrac{1}{4}y = 12$

$2x + y = 18$

6) $3x - 6y = 6$

$y = -2x$

Find an equation of the line that is **parallel** to the graph of each equation and passes through given point.

7) $y = 4x + 12$; $(2, 3)$

8) $y = \dfrac{1}{2}x - 6$; $(1, 5)$

9) $y = 5x - 7$; $(-1, -4)$

10) $y = \dfrac{2}{3}x - 18$; $(1, 3)$

Find an equation of the line that is **perpendicular** to the graph of each equation and passes through the given point.

11) $y = -x - 3$; $(2, -3)$

12) $\dfrac{1}{2}x - \dfrac{1}{3}y = 0$; $(-1, -4)$

13) $y + x = 9$; $(4, 8)$

14) $x = 2y - 5$; $(-1, -2)$

ADDING AND SUBTRACTING POLYNOMIALS

Monomial: x, y, z, ……. etc.(one term)

Binomial: 2x + y, x + y, a+b, ……. etc. (two terms)

Trinomial: ax^2 + bx + c, x + y + z, ……. etc. (three terms)

Like terms: like terms are terms that have the same variables with same power.

Example: 7x and 10x

Not Polynomials:

Examples:

$\dfrac{5}{x}$ Its not polynomials because variable can not be in denominator.

$\sqrt{x}$ Its not polynomials because variable can not be in radicals.

$x^{\frac{1}{3}}$ Its not polynomials because variable can not be in rational exponents.

x^{-5} Its not polynomials because variable can not be in negative power.

Adding Polynomials:

Example: add x^2 + 3x + 5 and $4x^2$ – 6x + 10.

Solution: x^2 + 3x + 5+ $4x^2$ – 6x + 10 Add like terms.

= $5x^2$ – 3x + 15

Subtracting Polynomials:

Example: Simplify $3x^2$ – 6x + 17 – (x^2 – 8x – 6).

Solution: First use distributive property for – (x^2 – 8x – 6) = $-x^2$ + 8x + 6

Now combine like terms: $3x^2$ – 6x + 17 – x^2 + 8x + 6

= $2x^2$ + 2x + 23

ADDING AND SUBTRACTING POLYNOMIALS EXERCISES

Add or subtract each of the following expressions.

1) $(x^2 - 6x) + (3x^2 - 8x)$

2) $(4x^2 - 3x) + (x^2 - 9x)$

3) $(x^2 - x + 3) + (x^2 - 8x + 10)$

4) $(x^2 - 3x + 15) + (x^2 - 10x)$

5) $(x^2 - 9x) + (5x^2 - 18x - 25)$

6) $\left(\dfrac{1}{2}x^2 + 2x + 5\right) + \left(\dfrac{3}{2}x^2 - 8\right)$

7) $(x^3 - 9x^2 + 8x - 1) + (x^2 - 18x)$

8) $(x^4 + x^3 + 7) + (x^2 - 12)$

9) $(x^3 - x^2 + 8x - 1) - (x^2 - x)$

10) $(x^2 + 7) - (x^2 - x - 1)$

11) $(-x^2 + x - 1) - (2x^2 - 3x + 4)$

12) $(x^4 + 7x + 5) - (x^2 - 10x + 3)$

13) $\left(\dfrac{1}{2}x^3 + 3\right) - (x^2 - 1)$

14) $\left(\dfrac{3}{2}x^2 + 2x + 5\right) - \left(\dfrac{3}{2}x^2 - 5x + 12\right)$

15) $(3x^2 + 3x^2) - (6x^2 - x)$

16) $(6x^2 - 2x) - (2x - 5x^2)$

17) $(12x - x^3) - (x^3 - 3x)$

18) $(x^4 - 4x) - (x - 2x^4)$

MULTIPLYING AND DIVIDING POLYNOMIALS

Exponent's rules:

- $x^a \times x^b = x^{a+b}$

- $\dfrac{x^a}{x^b} = x^{a-b}$

- $(x^a)^b = x^{a \times b}$

- $\dfrac{1}{x^a} = x^{-a}$

Multiplying Polynomials:

- Use foil method

- Combine like terms

Foil Method:

$$(ax + b)\ (cx + d) = ax \cdot cx + ax \cdot d + b \cdot cx + b \cdot d$$

Example: find the product of $x - 4$ and $2x - 7$

Solution: $(x - 4)(2x - 7)$

$= x(2x) - x(7) - 4(2x) + 4(7)$ Use foil method

$= 2x^2 - 7x - 8x + 28$ Combine like terms

$= 2x^2 - 15x + 28$

Dividing Polynomials:

- Divide the numerical coefficients

- Subtract exponents or like variables

Example: $\dfrac{12x^5}{6x^3} = \dfrac{12}{6} \times \dfrac{x^5}{x^3} = 2x^{5-3} = 2x^2$

MULTIPLYING AND DIVIDING POLYNOMIALS
EXERCISES

Find each product.

1) $(x - 1)(x + 1)$

2) $(x - 4)(x + 5)$

3) $(2x + 1)(3x - 5)$

4) $(x^2 - 2)(x^2 + 2)$

5) $\left(\frac{1}{2}x + 1\right)\left(\frac{1}{2}x - 1\right)$

6) $(x - 7)(x + 8)$

7) $(x)\left(\frac{1}{x} - 5\right)$

8) $(x)(x + 12)$

9) $(-4x)(3x - 5)$

10) $(5x^2)(x^2 + 2x - 1)$

Simplify each of following expression.

11) $\dfrac{25x^2y^3}{xy}$

12) $\dfrac{1x^7y^{12}}{2x^4y^5}$

13) $\dfrac{25x^{-8}y^{10}}{5x^{10}y^{-5}}$

14) $\dfrac{-33x^{20}y^{13}}{3x^8y^7}$

15) $\dfrac{4x^0y^0}{8x^1y^{-1}}$

16) $\dfrac{-2x^2y^3}{x^2y^3}$

17) $(5x^2y)(x^2 - y^2)$

18) $(5x^2y^4)(4x^2y^8)$

SPECIAL PRODUCT AND FACTORING POLYNOMIALS

Difference of Square:

$a^2 - b^2 = (a + b)(a - b)$

Examples:

Factor: $x^2 - 4^2 = (x - 2)(x + 2)$

Factor: $4x^2 - 25y^2 = (2x - 5y)(2x + 5y)$

Factor: $4x^3 - 16x$

$= 4x(x^2 - 4)$ First factor out the GCF, 4x

$= 4x(x - 2)(x + 2)$ Then factor the difference of squares

Difference of Cube:

$a^3 - b^3 = (a - b)(a^2 + 2ab + b^2)$

Examples:

Factor: $x^3 - 27 = x^3 - 3^3 = (x - 3)(x^2 + 6x + 9)$

Factor: $x^3 - 8 = x^3 - 2^3 = (x - 2)(x^2 + 4x + 4)$

Factoring by grouping:

Example:

Factor $x^3 - x^2 + 2x - 2$ by grouping.

Solution: $x^3 - x^2 + 2x - 2$

$= (x^3 - x^2) + (2x - 2)$

$= (x^3 - x^2) + (2x - 2)$

$= x^2(x - 1) + 2(x - 1)$

$= (x^2 + 2)(x - 1)$

SPECIAL PRODUCT AND FACTORING POLYNOMIALS EXERCISES

Factor each expression by factoring out the GCF.

1) $x^2 - 81x$

2) $16x^2 - 4x$

3) $8x^3 - 64x$

4) $25x^3 - 5x$

5) $27x^3 - 9x$

6) $x^3 + 2x^2 + x$

Factor each trinomial.

7) $x^2 - 5x + 6$

8) $x^2 + 5x + 6$

9) $x^2 - 11x + 30$

10) $x^2 - x - 20$

11) $x^2 + x - 72$

12) $x^2 + 2x - 80$

Factor each by grouping.

13) $6x^3 + 8x^2 + 3x + 4$

14) $x^3 + 3x^2 - x - 3$

15) $2x^3 + x^2 + 2x + 1$

16) $x^3 + 4x^2 - 3x + 2$

17) $x^3 + x^2 - x - 1$

18) $x^3 - 4x^2 - 8x + 32$

SIMPLIFYING RATIONAL EXPRESSIONS

To simplifying a rational expression:

Step 1: Factor the numerator and the denominator.

Step 2: Cancel out common factors in the numerator and denominator.

Step 3: Rewrite the remaining factors.

Example: Simplify $\dfrac{x^2-25}{x(x+5)}$

Solution:

$= \dfrac{x^2-25}{x+5}$ Factor the numerator and the denominator .

$= \dfrac{(x-5)\cancel{(x+5)}}{\cancel{(x+5)}}$ Cancel out common factors.

$= x-5$ Rewrite the remaining factors.

Example: Simplify $\dfrac{6-12x}{x^2-4}$

Solution:

$= \dfrac{6-12x}{x^2-4}$ Factor the numerator and the denominator.

$= \dfrac{-6\cancel{(x-2)}}{\cancel{(x-2)}(x+2)}$ Cancel like factors.

$= \dfrac{-6}{x-2}$ Rewrite the remaining factors.

Simplify each of the following rational expression.

1) $\dfrac{4-16x^2}{1+2x}$

2) $\dfrac{x-2}{x^2-4}$

3) $\dfrac{2x-6}{x-3}$

4) $\dfrac{x^2-9}{x^4-81}$

5) $\dfrac{9x^2-25}{3x-5}$

6) $\dfrac{7x-28}{x-4}$

7) $\dfrac{9x^2}{3x-6}$

8) $\dfrac{x^2-6x}{x^2-36}$

9) $\dfrac{(x-2)(x+4)}{(x+2)(x+4)}$

10) $\dfrac{x^2-x-6}{x^2-7x+12}$

11) $\dfrac{x^2-3}{x-\sqrt{3}}$

12) $\dfrac{x^4-x^2}{x^2-1}$

13) $\dfrac{x^3+x^2+2x}{x+1}$

14) $\dfrac{x-5}{x^2-25}$

15) $\dfrac{4x^2-1}{2x+1}$

16) $\dfrac{2x^2+7x+3}{x^2+2x-3}$

17) $\dfrac{x^2-8}{x-2\sqrt{2}}$

18) $\dfrac{x^2-25}{x^3-125}$

ADDING AND SUBTRACTING RATIONAL EXPRESSIONS

A rational expression is a fraction that can be written as a ratio of two polynomials.

When the denominator is zero, the rational expression is undefined.

Adding Rational Expression: When you add rational expressions, if they have same denominator, you add the numerators while keeping the denominator the same.

Subtracting Rational Expression: When you subtract rational expressions that have the same denominators, you subtract only the numerators and keep the denominator the same.

To adding or subtracting rational expressions:

- Find least common denominator.
- Add or subtract like terms in numerators.
- Simplify the expression as needed.

Example: Simplify $\dfrac{x}{x+1} + \dfrac{2x}{x+1}$

Solution: $\dfrac{x}{x+1} + \dfrac{2x}{x+1} = \dfrac{x+2x}{x+1} = \dfrac{3x}{x+1}$

Example: Simplify $\dfrac{1}{x+3} + \dfrac{x}{x-1}$

Solution: $\dfrac{1}{x+3} + \dfrac{x}{x-1}$

$$= \dfrac{1(x-1)}{(x+3)(x-1)} + \dfrac{x(x+3)}{(x-1)(x+3)}$$

$$= \dfrac{x-1}{(x+3)(x-1)} + \dfrac{x^2+3}{(x-1)(x+3)}$$

$$= \dfrac{x^2+x+2}{(x+3)(x-1)}$$

ADDING AND SUBTRACTING RATIONAL EXPRESSIONS
EXERCISES

Simplify each of the following expression.

1) $\dfrac{x}{x+2} + \dfrac{1}{x}$

2) $\dfrac{x}{x-2} - \dfrac{1}{x}$

3) $\dfrac{2x}{x-1} + \dfrac{3x}{x+1}$

4) $\dfrac{x+2}{x-2} - \dfrac{x-1}{x}$

5) $\dfrac{x+2}{x-2} + \dfrac{x+3}{x-3}$

6) $\dfrac{x^2-4}{x-2} + \dfrac{x^2-25}{x-5}$

7) $\dfrac{x^2-1}{x+1} + \dfrac{x-3}{x-3}$

8) $\dfrac{x^2+1}{x-1} - \dfrac{x^2-1}{x+1}$

9) $\dfrac{x^2+2x+1}{x+1} + \dfrac{x^2+6x+9}{x+3}$

10) $\dfrac{x^2-x-12}{x-4} - \dfrac{x^2+2x-3}{x+3}$

11) $\dfrac{1}{x-4} + \dfrac{1}{x+3}$

12) $\dfrac{x+3}{x-4} - \dfrac{3x-1}{x^2-4x}$

13) $\dfrac{4}{x-4} + \dfrac{5}{x-5}$

14) $\dfrac{x^2-81}{x-9} - \dfrac{x^2-16}{x-4}$

15) $\dfrac{x-7}{x} + \dfrac{x-5}{2x}$

16) $\dfrac{3x}{x-3} - \dfrac{x}{x+3}$

17) $\dfrac{x^3+y^3}{x^2-xy+y^2} + \dfrac{2x+2y}{x+y}$

18) $\dfrac{x+1}{x^2-1} - \dfrac{x-1}{x^2+1}$

To multiply rational expressions:

- Complete factor of each numerator and each denominator.
- Cancel common factors.
- Simplify as needed.

Example: Simplify $\dfrac{x-3}{x-4} \cdot \dfrac{x^2-16}{x-3}$.

Solution:

$$= \frac{x-3}{x-4} \cdot \frac{x^2-16}{x-3} \qquad \text{Complete factors.}$$

$$= \frac{x\cancel{(x-3)}}{\cancel{x-4}} \cdot \frac{\cancel{(x-4)}(x+4)}{\cancel{x-3}} \qquad \text{Cancel common factors.}$$

$$= x\,(x+4) \qquad \text{Simplify}$$

$$= x^2 + 4x$$

To divide rational expressions:

- Change division sign to multiplication sign and then flip the second rational expressions.
- Complete factor of each numerator and each denominator.
- Cancel out common factors.
- Simplify as needed.

Example: Simplify $\dfrac{x^2-9}{x+3} \div \dfrac{x-3}{x+3}$

Solution:

$$= \frac{x^2-9}{x+3} \div \frac{x-3}{x+3} \qquad \text{Multiply by the reciprocal.}$$

$$= \frac{\cancel{(x-3)}(x+3)}{\cancel{(x+3)}} \cdot \frac{(x+3)}{\cancel{(x-3)}} \qquad \text{Cancel common factors.}$$

$$= x + 3$$

MULTIPLYING AND DIVIDING RATIONAL EXPRESSIONS EXERCISES

Simplify each of the following expression.

1) $\dfrac{x}{x+2} \cdot \dfrac{x^2-4}{x}$

2) $\dfrac{x-3}{x-4} \cdot \dfrac{x^2-16}{x^2-9}$

3) $\dfrac{7x}{12} \cdot \dfrac{36x}{14x^2}$

4) $\dfrac{3x+30}{21x} \cdot \dfrac{7x+14}{x+10}$

5) $\dfrac{1}{x-8} \cdot \dfrac{x^2-64}{x+8}$

6) $\dfrac{2x+3}{x-7} \cdot \dfrac{2x-14}{4x+6}$

7) $\dfrac{x^3-y^3}{x^2+xy+y^2} \cdot \dfrac{x}{x-y}$

8) $\dfrac{x^2-100}{x-10} \cdot \dfrac{x^2-25}{x-5}$

9) $\dfrac{4x}{x-6} \div \dfrac{4x}{2x-12}$

10) $\dfrac{x^2-64}{x-8} \div \dfrac{x^2-49}{x+7}$

11) $\dfrac{x^2+2x+1}{x+1} \div \dfrac{x^2+6x+9}{x+3}$

12) $\dfrac{x^2-8x+15}{x^2-9} \div \dfrac{x^2-4x-5}{x^2+3x}$

13) $\dfrac{x^2-y^2}{x^2+xy} \div \dfrac{x^2-xy}{xy+x}$

14) $\dfrac{x^2-y^2}{x-y} \div \dfrac{x^2+y^2}{x+y}$

15) $\dfrac{x^2+4x+16}{x+4} \div \dfrac{x^2-6x+9}{x-3}$

16) $\dfrac{y}{y-1} \div \dfrac{y^2}{y^3-y}$

17) $\dfrac{3x^2+5x-2}{3x-1} \div \dfrac{x+2}{5}$

18) $\dfrac{12x^2-14x-6}{3x+1} \div \dfrac{4x-6}{x^2}$

COMPLEX FRACTIONS

Complex Fraction is a fraction that contains another fraction in its numerator, denominator or both. For simplify complex fraction divide the fraction in the numerator by the fraction in the denominator.

Rule: $\dfrac{\frac{a}{b}}{\frac{c}{d}} = \dfrac{a}{b} \cdot \dfrac{d}{c} = \dfrac{ad}{bc}$

Example: $\dfrac{\frac{1}{2}}{\frac{3}{4}} = \dfrac{1}{2} \cdot \dfrac{4}{3} = \dfrac{4}{6} = \dfrac{2}{3}$

Example: $\dfrac{\frac{1}{6}}{\frac{1}{5}} = \dfrac{1}{6} \cdot \dfrac{5}{1} = \dfrac{5}{6}$

Example: $\dfrac{\frac{x+1}{4}}{\frac{x-7}{3}} = \dfrac{x+1}{4} \cdot \dfrac{3}{x-7} = \dfrac{3(x+1)}{4(x-7)} = \dfrac{3x+3}{4x-28}$

Example: $\dfrac{\frac{x-1}{5}}{\frac{x}{4} - \frac{x+2}{3}} = \dfrac{\frac{x-1}{5}}{12\left(\frac{x}{4} - \frac{x+2}{3}\right)}$

$= \dfrac{\frac{x-1}{5}}{\overset{3}{\cancel{12}} \cdot \frac{x}{\underset{1}{\cancel{4}}} - \overset{4}{\cancel{12}} \cdot \frac{x+2}{\underset{1}{\cancel{3}}}} = \dfrac{\frac{x-1}{5}}{3x - 4(x+2)} = \dfrac{\frac{x-1}{5}}{3x - 4x - 8} = \dfrac{\frac{x-1}{5}}{-x-8}$

$= \dfrac{x-1}{5} \cdot \dfrac{1}{(-x-8)} = \dfrac{x-1}{-5x-40}$

COMPLEX FRACTIONS EXERCISES

Simplify each of following expression.

1) $\dfrac{\frac{1}{3}}{\frac{3}{7}}$

2) $\dfrac{\frac{x}{y}}{\frac{a}{b}}$

3) $\dfrac{\frac{x-1}{6}}{\frac{x-2}{18}}$

4) $\dfrac{\frac{x+1}{24}}{\frac{x-4}{48}}$

5) $\dfrac{\frac{x-3}{5}}{\frac{x}{2}-\frac{x}{3}}$

6) $\dfrac{\frac{x-2}{6}-\frac{x+2}{3}}{\frac{x}{4}}$

7) $\dfrac{x}{\frac{x}{3}-\frac{x}{5}}$

8) $\dfrac{1+\frac{1}{x}}{1-\frac{2}{x}}$

9) $\dfrac{\frac{a}{b}}{x}$

10) $\dfrac{x}{\frac{3}{x}-\frac{5}{x}}$

11) $\dfrac{\frac{2}{1\frac{1}{x}}}{3\frac{3}{x}}$

12) $\dfrac{\frac{x}{5}}{\frac{x}{4}-\frac{x}{8}}$

13) $\dfrac{\frac{2}{a}}{\frac{a}{6}}$

14) $\dfrac{\frac{x}{5}}{\frac{10}{x}}$

FUNCTION NOTATION, DOMAIN, RANGE AND INVERSE FUNCTIONS

Key Notes:

Function: If every element of A is paired with the elements of B at least once and $A \neq 0$ and $B \neq 0$, this correlation is called a function.

Inverse Function: Let f be a one to one function with domain of x and range of y. The inverse of this function is $f^{-1}\{y\} = x$.

$f(x) = y \Leftrightarrow f^{-1}(y) = x$

Function Notation

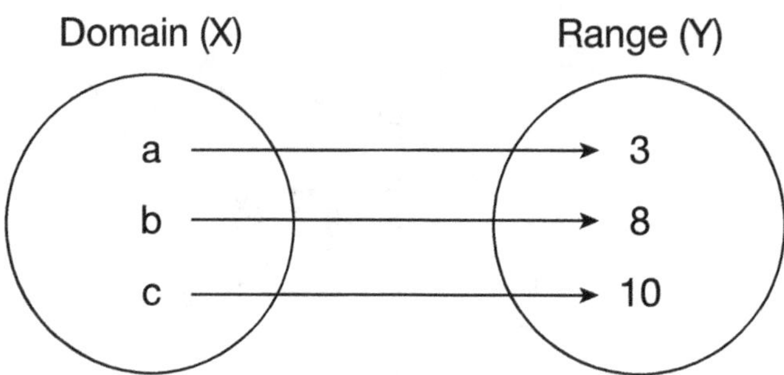

Example:

Not a Function: If the domain (x) is repeating, it's not a function. You do not need to check the range (y). It does not matter if the range repeats.

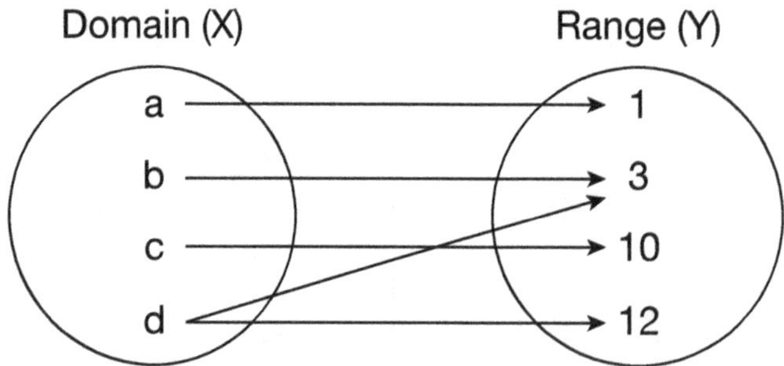

Examples:

(a, 1), (b, 3), (c, 5) is a function because domain elements are not repeating.

(a, 1), (b, 3), (c, 5),(a,10) is not a function because domain elements are repeating.

FUNCTION NOTATION, DOMAIN, RANGE AND INVERSE FUNCTIONS EXERCISES

For each of following points, state whether or not it is a function.

1) (2, 3), (5, 3), (6, 7)

2) (1, 3), (2, 4), (3, 5), (1,7)

Evaluate each of following function.

3) $f(x) = 2x-8$, find $f(-2)$

4) $f(x) = x^2 - 5$, find $f(3)$

5) $f(x) = 2x^2 + 3$, find $f\left(\dfrac{1}{2}\right)$

6) $f(x) = x^2 + 6$, find $f(7)$

7) $f(x) = 2x^2 + 2x - 5$, find $f(-3)$

8) $f(x) = x^2 - 10x + 12$, find $f\left(-\dfrac{1}{2}\right)$

9) $h(x) = \dfrac{4x-1}{3}$, find $h(-2)$

10) $h(x) = \dfrac{x^2-6x+12}{3}$, find $h(3)$

11) $g(x) = 2x^2 + 4x + 5$, find $h\left(\dfrac{x}{2}\right)$

12) $g(x) = x^2 - 6x$, find $g(3x)$

13) $h(x) = x^2 + 12$, find $h(\sqrt{x})$

14) $g(x) = 3x - 5$, find $g(-3x)$

Find the Inverse of each of the following functions.

15) $f(x) = x + 4$

16) $f(x) = 2x - 54$

17) $f(x) = x^2 - 3$

18) $f(x) = \sqrt{x-6}$

1) Adding: $(f + g)(x) = f(x) + g(x)$

Example: Let $f(x) = 4x-2$ and $g(x) = 2x - 1$. Then $(4x - 2)+(2x - 1) = 6x - 3$

2) Subtracting: $(f - g)(x) = f(x) - g(x)$

Example: Let $f(x) = 4x - 2$ and $g(x) = 2x - 1$, then $(4x - 2) - (2x - 1) = 2x - 1$

3) Multiplying: $(f \cdot g)(x) = f(x) \cdot g(x)$

Example: Let $f(x) = 4x - 2$ and $g(x) = 2x - 1$, then $(4x - 2) \cdot (2x - 1) = 8x^2 - 8x + 2$

4) Dividing: $(f \div g)(x) = f(x) \div g(x)$

Example: Let $f(x) = 4x - 2$ and $g(x) = 2x - 1$,

$$\text{then } (4x - 2) \div (2x - 1) = \frac{4x - 2}{2x - 1} = \frac{2(2x - 1)}{2x - 1} = 2$$

OPERATIONS WITH FUNCTIONS AND COMPOSITION OF FUNCTIONS EXERCISES

Adding and Subtracting Functions.

Let $f(x) = -3x + 9$ and $g(x) = 2x + 5$

1) Find $f(x) + g(x)$

2) Find $f(x) - g(x)$

Let $f(x) = x^2 + 3x + 9$ and $g(x) = x + 12$

3) Find $f(x) + 2g(x)$

4) Find $3f(x) - g(x)$

Find each value or expression

Let $f(x) = x + 4$ and $g(x) = 3x + 10$

5) Find $f(g(x))$

6) Find $g(f(x))$

Let $f(x) = 5x$ and $g(x) = 6x - 15$

7) Find $f(g(x))$

8) Find $g(f(x))$

Let $f(x) = 2x^2 + 1$ and $g(x) = x - 3$

9) Find $f(g(3))$

10) Find $g(f(5))$

Let $f(x) = 2x^2 + x + 3$ and $g(x) = 4x^2 + 5x - 12$

11) Find $f(g(-1))$

12) Find $g(f(-3))$

Multiplying and Dividing Functions.

Let $f(x) = x^2 - x$ and $g(x) = x^2 + x - 2$

13) Find $f(x) \cdot g(x)$

14) Find $\dfrac{f(x)}{g(x)}$

EXPONENTIAL FUNCTIONS

$y = a \cdot b^x$, where $a \neq 0$, $b > 0$ and x is a real number.

Example: Solve $16^{2x-5} = 32^{x-3}$

Solution:

$16^{2x-5} = 32^{x-3}$,

$2^{4(2x-5)} = 2^{5(x-3)}$, Since bases are same powers must be same as well.

$4(2x - 5) = 5(x - 3)$

$8x - 20 = 5x - 15$,

$8x - 5x = 20 - 15$

$3x = 5$, $x = \dfrac{5}{3}$

Example: If $f(x) = \dfrac{2^{x-6}}{8}$, then find $f(-1)$.

Solution: If $f(x) = \dfrac{2^{x-6}}{8}$, Then

$f(-1) = \dfrac{2^{-1-6}}{2^3} = \dfrac{2^{-7}}{2^3} = 2^{-7-3} = 2^{-10} = \dfrac{1}{2^{10}}$

Exponential Growth

$y = a(1 + r)^t$

Exponential Decay

$y = a(1 - r)^t$

Example: Turkey's currently population is approximately 81 million. If Turkey population were growing at a rate of 3% per year, what equation would model their population growth?

Solution:

$a = 81$ million

$r = 3\% = 0.03$

$f(t) = a(1+r)^t$, $= 81(1 + 0.03)^t = 81(1.03)^t$

EXPONENTIAL FUNCTIONS EXERCISES

1) If $f(x) = 2^x$, then find f(–2).

2) If $f(x) = 3^{x-2}$, then find f(4).

3) If $f(x) = \dfrac{2^{x-5}}{4}$, then find f(–1).

4) Graph $f(x) = 3^{x-1}$

5) Graph $f(x) = 2^x + 1$

6) Solve $3^{x-1} = 243$

7) Solve $2 \cdot (2)^{x+1} = 32$

8) Solve $9^{2x-3} = 81^{3x-8}$

9) Solve $\left(\dfrac{3}{2}\right)^{x-2} = \left(\dfrac{9}{4}\right)^{3x+2}$

10) Solve $8^{2x-5} = 64^{3x-2}$

11) For following table, determinate if the relation is exponential function of x. If so, write its equation.

x	y
0	1
1	5
2	25
3	125
4	625

12) Covid–19 spread exponentially. In fact if you start with 1 person infected with the covid–19 virus, after one day a total of 20 people, after two days a total of 400 people, after three days a total of 8000 people would be covid–19. What equation models this situation?

TRIGONOMETRIC FUNCTIONS

Trigonometric Values of Special Angle

Degrees	0°	30°	45°	60°	90°	180°	360°
Radians	0	$\frac{\pi}{6}$	$\frac{\pi}{4}$	$\frac{\pi}{3}$	$\frac{\pi}{2}$	π	2π
sin∅	0	$\frac{1}{2}$	$\frac{1}{\sqrt{2}}$	$\frac{\sqrt{3}}{2}$	1	0	0
cos∅	1	$\frac{\sqrt{3}}{2}$	$\frac{1}{\sqrt{2}}$	$\frac{1}{2}$	0	–1	1
tan∅	0	$\frac{1}{\sqrt{3}}$	1	$\sqrt{3}$	Und	0	0
cot∅	Und	$\sqrt{3}$	1		0	Und	Und

Sinus and cosines Property

- ✓ sin(x)= cos(90°–x)
- ✓ cos(x)= sin(90°–x)

Example: Evaluate sin120°.

Solution:

sin(x) = cos(90°–120°), then sin120° = –cos(30°) = $\frac{-\sqrt{3}}{2}$

Coterminal angles

Coterminal angles are equal angle. Finding coterminal an angle, adding or subtracting 360° or 2π to each angle.

Example: Find a positive and a negative angle coterminal to angle 55°.

Solution:

55° + 360° = 415° or 55° – 360° = –305°

TRIGONOMETRIC FUNCTIONS

Formulas for right triangles

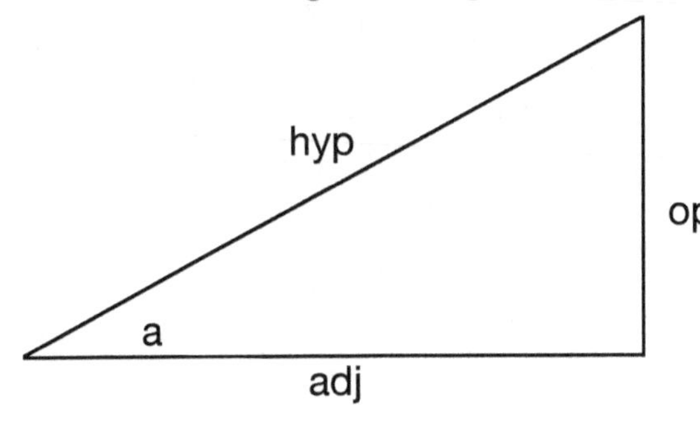

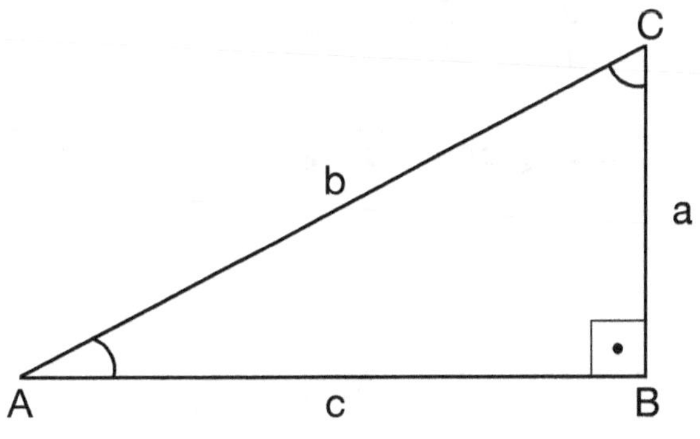

$$\sin\alpha = \frac{\text{opposite}}{\text{hypotenuse}}$$

$$\sin\widehat{A} = \frac{|BC|}{|AC|} = \frac{a}{b}$$

$$\sin\widehat{C} = \frac{|AB|}{|AC|} = \frac{c}{b}$$

$$\cos\alpha = \frac{\text{adjacent}}{\text{hypotenuse}}$$

$$\cos\widehat{A} = \frac{|AB|}{|AC|} = \frac{c}{b}$$

$$\cos\widehat{C} = \frac{|BC|}{|AC|} = \frac{a}{b}$$

$$\tan\alpha = \frac{\text{opposite}}{\text{adjacent}}$$

$$\tan\widehat{A} = \frac{|BC|}{|AB|} = \frac{a}{c}$$

$$\tan\widehat{C} = \frac{|AB|}{|BC|} = \frac{c}{a}$$

$$\cot\alpha = \frac{\text{adjacent}}{\text{opposite}}$$

$$\cot\widehat{A} = \frac{|AB|}{|BC|} = \frac{c}{a}$$

$$\cot\widehat{C} = \frac{|BC|}{|AB|} = \frac{a}{c}$$

Example: From the following right triangle, find cos(30°).

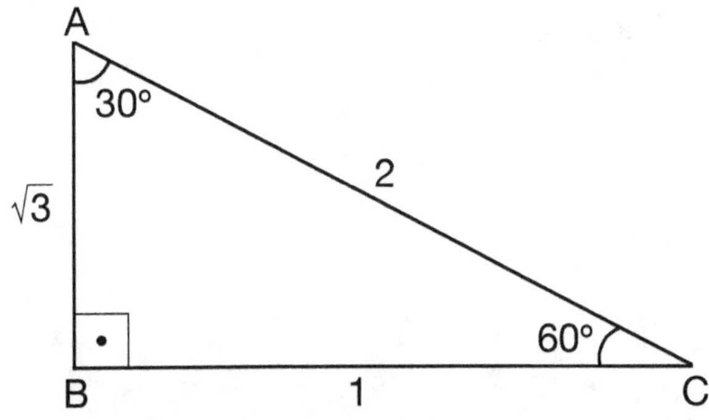

TRIGONOMETRIC FUNCTIONS

Solution: $\cos(30°) = \dfrac{adj}{hyp} = \dfrac{\sqrt{3}}{2}$

Convert Radians to Degrees

D: Degrees

R: Radians

$$\frac{D}{180} = \frac{R}{\pi}$$

Example: Convert 150 degrees to radians.

Solution: From $\dfrac{D}{180°} = \dfrac{R}{\pi}$, then $\dfrac{150}{180°} = \dfrac{R}{\pi}$ (cross–multiply) $R = \dfrac{150\pi}{180°} = \dfrac{15\pi}{18} = \dfrac{5\pi}{6}$

Length of an Arc formula:

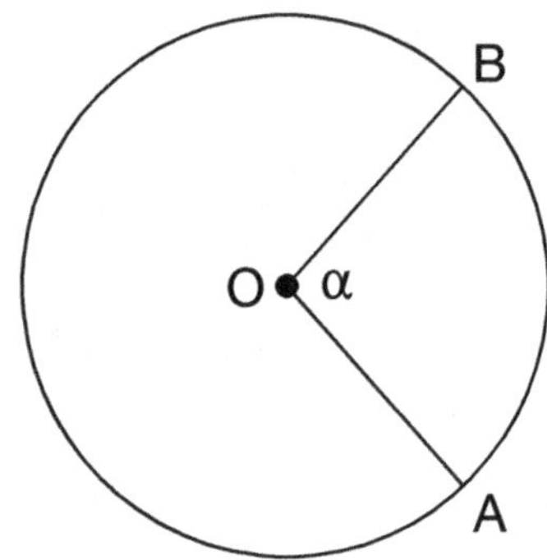

$$\text{Arc}AB = 2\pi r \frac{\alpha}{360°}$$

Area of the Sector:

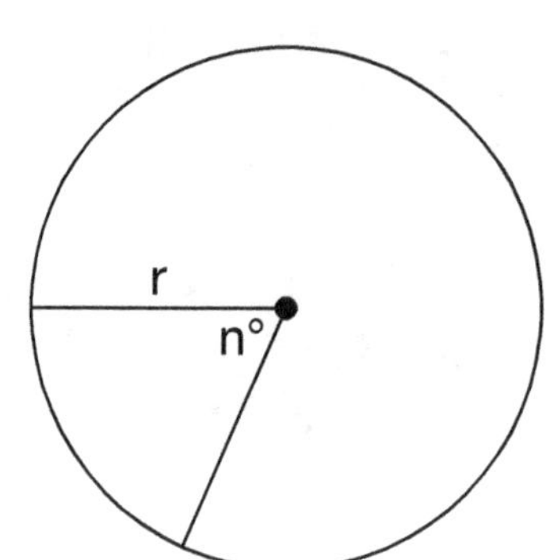

$$A = \frac{n°}{360°}\pi r^2$$

Example: If $r = 5$ and $\theta = 120°$, then find the length of the Arc. (Find your answer in terms of π)

Solution: Length of the Arc $= \dfrac{2\pi r\theta}{360°} = \dfrac{2.5.120.\pi}{360°} = \dfrac{10\pi}{3}$

Example: If $r = 4$ and $\theta = 60°$, then find the area of sector. (Find your answer in terms of π)

Solution: Length of the Arc $= \dfrac{\pi r^2\theta}{360°}$

$= \dfrac{\pi 4^2 \cdot 60°}{360°} = \dfrac{16\pi \cdot 60°}{360°} = \dfrac{16\pi}{6} = \dfrac{8\pi}{3}$

TRIGONOMETRIC FUNCTIONS EXERCISES

1) Evaluate sin120°.

2) Evaluate cos135°.

3) Find a positive and a negative angle coterminal to angle 25°.

4) Find a positive and a negative angle coterminal to angle $\frac{\pi}{2}$

5) From the following right triangle, find sin(30°).

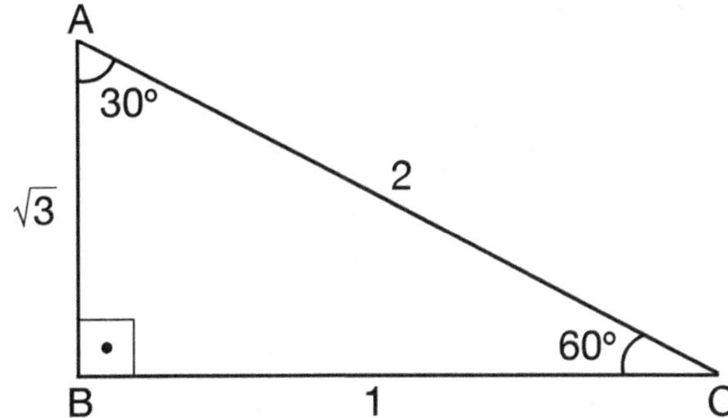

6) Convert 200 degrees to radians.

7) Convert $\frac{\pi}{5}$ radians degrees.

9) If r = 10 and θ = 90°, then find the area of sector. (Find your answer in terms of π)

10) If 0° < x 90° and sinx = $\frac{3}{5}$, then find cotx?

11) In a right triangle, one angle measures x, where cosx° = $\frac{5}{13}$.

What is the tan (90° – x°)?

SIMPLIFYING RADICAL EXPRESSIONS

Index $\longrightarrow \sqrt[n]{X} \longleftarrow$ Radical Sign

$\uparrow$

Radicand

Note: If the index not shown, it is assumed to be 2.

Example:

Index $\longrightarrow \sqrt[2]{7} \longleftarrow$ Radical Sign

$\uparrow$

Radicand

Square Roots of Perfect Squares Examples:

$$\sqrt{1} = \sqrt[2]{1^2} = 1, \qquad \sqrt{4} = \sqrt[2]{2^2} = 2, \qquad \sqrt{9} = \sqrt[2]{3^2} = 3$$

$$\sqrt{16} = \sqrt[2]{4^2} = 4, \qquad \sqrt{25} = \sqrt[2]{5^2} = 5, \qquad \sqrt{36} = \sqrt[2]{6^2} = 6$$

$$\sqrt{81} = \sqrt[2]{9^2} = 9, \qquad \sqrt{144} = \sqrt[2]{12^2} = 12, \qquad \sqrt{225} = \sqrt[2]{15^2} = 15$$

Simplifying Radicals Product Rule:

$$\sqrt[n]{a \times b} = \sqrt[n]{a} \times \sqrt[n]{b}$$

Example: $\sqrt[2]{20} = \sqrt[2]{4 \times 5} = \sqrt[2]{4} \times \sqrt[2]{5} = \sqrt[2]{2^2} \times \sqrt[2]{5} = 2\sqrt[2]{5} = 2\sqrt{5}$

Example: $\sqrt[2]{60} = \sqrt[2]{4 \times 15} = \sqrt[2]{4} \times \sqrt[2]{15} = \sqrt[2]{2^2} \times \sqrt[2]{15} = 2\sqrt[2]{15} = 2\sqrt{15}$

Example: $\sqrt[2]{80} = \sqrt[2]{16 \times 5} = \sqrt[2]{16} \times \sqrt[2]{5} = \sqrt[2]{4^2} \times \sqrt[2]{5} = 4\sqrt[2]{5} = 4\sqrt{5}$

SIMPLIFYING RADICAL EXPRESSIONS
EXERCISES

Simplify each of the following radical.

1) $\sqrt{169}$

2) $\sqrt{400}$

3) $\sqrt{144}$

4) $\sqrt{10000}$

5) $\sqrt{20}$

6) $\sqrt{1000}$

7) $\sqrt{x^4}$

8) $\sqrt{y^{16}}$

9) $\sqrt{25x^4}$

10) $\sqrt{36y^6}$

11) $\sqrt{144x^4y^{10}}$

12) $\sqrt{64x^8y^6}$

13) $\sqrt{\dfrac{1}{25}x^4y^8}$

14) $\sqrt{400x^2y^2z^2}$

15) $\sqrt{20x^2y^3}$

16) $\sqrt{\dfrac{18x^8}{3x^3}}$

17) $\sqrt{81x^{12}y^{18}z^{16}}$

18) $\sqrt{\dfrac{64x^6y^8}{16xy}}$

ADDING AND SUBTRACTING RADICAL EXPRESSIONS

- Radical expressions can be added or subtracted only if they are same index and same radicals.

Like Radicals

$2\sqrt{3}$, and $7\sqrt{3}$

$2\sqrt{x}$, and $3\sqrt{x}$

Unlike Radicals

$2\sqrt{3}$, and $6\sqrt{5}$

$5\sqrt{x}$, and $7\sqrt{y}$

Example: Simplify $2\sqrt{3} + 6\sqrt{3} = 8\sqrt{3}$

Example: Simplify $7\sqrt{3} - 4\sqrt{3} = 3\sqrt{3}$

Example: Simplify $3\sqrt{x} + 2\sqrt{x} = 5\sqrt{x}$

Example: Simplify $7\sqrt{a} - 4\sqrt{a} = 3\sqrt{a}$

Example: Simplify $\sqrt{18} - \sqrt{8}$.

Solution: $\sqrt{9 \times 2} - \sqrt{4 \times 2}$ Perfect square factor.

$= \sqrt{9} \times \sqrt{2} - \sqrt{4} \times \sqrt{2}$ Use product property.

$= 3\sqrt{2} - 2\sqrt{2}$ Simplify.

$= \sqrt{2}$

ADDING AND SUBTRACTING RADICAL EXPRESSIONS EXERCISES

Simplify each of the following radicals.

1) $2\sqrt{2} + 3\sqrt{2}$

2) $6\sqrt{3} + 9\sqrt{3}$

3) $7\sqrt{2} - 3\sqrt{2}$

4) $12\sqrt{5} - 7\sqrt{5}$

5) $\sqrt{20} + \sqrt{18}$

6) $\sqrt{24} + \sqrt{72}$

7) $\sqrt{63} - \sqrt{28}$

8) $\sqrt{48} - \sqrt{12}$

9) $6\sqrt{2} + \sqrt{32}$

10) $5\sqrt{4} + 3\sqrt{36}$

11) $18\sqrt{2} - \sqrt{98}$

12) $48\sqrt{2} - 3\sqrt{128}$

13) $x\sqrt{2} + 2x\sqrt{2}$

14) $4y\sqrt{2} + 3\sqrt{72y^2}$

15) $2xy\sqrt{2} - xy\sqrt{18}$

16) $5\sqrt{2} - 3\sqrt{2} + 8\sqrt{2} - \sqrt{2}$

17) $\sqrt{48xy} - \sqrt{75xy}$

18) $9\sqrt{x} - 3\sqrt{y} + 5\sqrt{x} - \sqrt{y}$

MULTIPLYING AND DIVIDING RADICAL EXPRESSIONS

- Use distributive property and product rule for multiplying radicals.

Product Rule:

$$\sqrt[n]{a} \times \sqrt[n]{b} = \sqrt[n]{a \times b}$$

Example: $\sqrt[2]{2} \times \sqrt[2]{3} = \sqrt[2]{2 \times 3} = \sqrt{6}$

Distributive Property Rule:

$$\sqrt{a}\,(b + \sqrt{b}) = \sqrt{a} \times b + \sqrt{a} \times \sqrt{b}$$

Example: Simplify $\sqrt{2}\,(3 + \sqrt{10})$.

Solution:

$= \sqrt{2}\,(3 + \sqrt{10})$ Use distributive rule.

$= 3 \times \sqrt{2} + \sqrt{2} \times \sqrt{10}$ Use product property.

$= 3\sqrt{2} + \sqrt{20}$ Perfect square factor.

$= 3\sqrt{2} + \sqrt{4 \times 5}$

$= 3\sqrt{2} + 2\sqrt{5}$

Quotient Property:

$$\frac{\sqrt[n]{a}}{\sqrt[n]{b}} = \sqrt[n]{\frac{a}{b}}$$

Example: $\dfrac{\sqrt[2]{5}}{\sqrt[2]{6}} = \sqrt[2]{\dfrac{5}{6}}$

Note: If the denominator has a radical number, we need to multiply top and bottom by the radical in the denominator to remove radicals.

Example: Simplify $\dfrac{\sqrt{3}}{\sqrt{6}}$

Solution: $\dfrac{\sqrt{3}}{\sqrt{6}} \times \dfrac{\sqrt{6}}{\sqrt{6}} = \dfrac{\sqrt{18}}{\sqrt{36}} = \dfrac{\sqrt{9 \times 2}}{6 \times 6} = \dfrac{3\sqrt{2}}{6} = \dfrac{\sqrt{2}}{2}$

MULTIPLYING AND DIVIDING RADICAL EXPRESSIONS EXERCISES

Simplify each of the following radicals.

1) $\sqrt{3}\,(2 + \sqrt{8})$

2) $\sqrt{5}\,(3 + \sqrt{2})$

3) $\sqrt{2}\,(5 - \sqrt{7})$

4) $\sqrt{10}\,(4 - \sqrt{10})$

5) $\sqrt{3}\,(\sqrt{6} + \sqrt{6})$

6) $\sqrt{8}\,(\sqrt{7} + \sqrt{7})$

7) $(\sqrt{2} + \sqrt{3})(\sqrt{2} - \sqrt{3})$

8) $(\sqrt{a} + \sqrt{b})(\sqrt{a} - \sqrt{b})$

9) $\sqrt{\dfrac{18}{x^2}}$

10) $\sqrt{\dfrac{24x^4 y^8}{6x^3 y^2}}$

11) $\dfrac{\sqrt{12}}{\sqrt{2}}$

12) $\dfrac{\sqrt{8}}{\sqrt{3}}$

13) $\dfrac{\sqrt{48}}{3\sqrt{2}}$

14) $\dfrac{\sqrt{72}}{6\sqrt{3}}$

15) $\dfrac{\sqrt{2} + \sqrt{3}}{\sqrt{2} - \sqrt{3}}$

16) $\dfrac{\sqrt{5} - \sqrt{7}}{\sqrt{5} + \sqrt{7}}$

17) $\dfrac{\sqrt{x} + \sqrt{y}}{\sqrt{x} - \sqrt{y}}$

18) $\dfrac{\sqrt{6} + \sqrt{7}}{2\sqrt{6} - \sqrt{7}}$

- If a and b are the rational number and x and y are real numbers then,

✓ $a^x \times a^y = a^{x+y}$

Example: $2^4 \times 2^5 = 2^{4+5} = 2^9$

✓ $(a^x)^y = a^{x \times y}$

Example:

$(a^3)^5 = a^{3 \times 5} = a^{30}$

✓ $a^0 = 1$

Examples: $5^0 = 1$, $100^0 = 1$, $101^0 = 1$

✓ $a^{-n} = \dfrac{1}{a^n}$

Example: $a^{-3} = \dfrac{1}{a^3}$

Example: $5^{-3} = \dfrac{1}{5^3}$

✓ $a^{-\frac{1}{n}} = \dfrac{1}{\sqrt[n]{a}}$ When $a \neq 0$

Example: $a^{-\frac{1}{4}} = \dfrac{1}{\sqrt[4]{a}}$

Example: $6^{-\frac{1}{5}} = \dfrac{1}{\sqrt[5]{6}}$

Example: $32^{-\frac{1}{4}} = (2^5)^{-\frac{1}{4}} = 2^{-\frac{5}{4}} = \dfrac{1}{\sqrt[4]{2^5}} = \dfrac{1}{2\sqrt[4]{2^1}}$

Example: $64^{\frac{1}{3}} = (2^6)^{\frac{1}{3}} = (2)^{\frac{6}{3}} = 2^2 = 4$

RATIONAL EXPONENTS EXERCISES

Evaluate each of the following (No CALCULATOR!)

1) $6^4 \times 6^5$

2) $3^{-4} \times 3^{-7}$

3) $(2^3)^{-4}$

4) $(a^{-7})^5$

5) $27^{-\frac{1}{3}}$

6) $81^{-\frac{1}{5}}$

7) $\left(\dfrac{1}{27}\right)^{-\frac{1}{3}}$

8) $\left(\dfrac{1}{4}\right)^{-\frac{1}{2}}$

9) $4^{-\frac{1}{3}}$

10) $7^{-\frac{1}{5}}$

11) $x^{-\frac{1}{5}}$

12) 125°

13) $(64x^8y^4)^{-\frac{1}{3}}$

14) $(x^{-10}y^5)^{-\frac{1}{5}}$

15) $\dfrac{(3x^3y^6)^{-3}}{y^{-7}}$

16) $\dfrac{(2x^4y^{12}z^{16})^4}{6x^6y^{20}y^{-7}}$

17) $\dfrac{15(x^2y^{-3})^{-6}}{(3^{-1}y)^{-3}}$

18) $\left(\dfrac{x^3}{\sqrt{x^4}}\right)^{-\frac{1}{2}}$

IMAGINARY NUMBERS AND COMPLEX NUMBERS

An imaginary number is a complex number that can be written as a real number multiplied by the imaginary number. It's defined by $i^2 = -1$

Powers of i:

$$i^1 = i \qquad i^2 = -1 \qquad i^3 = -i \qquad i^4 = 1$$
$$i^5 = i \qquad i^6 = -1 \qquad i^7 = -i \qquad i^8 = 1$$
$$i^9 = i \qquad i^{10} = -1 \qquad i^{11} = -i \qquad i^{12} = 1$$
$$i^2 = -1 \text{ or } i = \sqrt{-1}$$

Complex Numbers

$$a + bi$$

Real Numbers   Imaginary Numbers

Complex number in standard form: $a + bi$

Example: Complex number: $-3 + \sqrt{-4}$

$$-3 + \sqrt{-4} = -3 + \sqrt{(-1(2^2))} = -3 + 2\sqrt{(-1)} = -3 + 2\sqrt{(i^2)} = -3 + 2i$$

Standard form: $-3 + 2i$

Adding and Subtracting Complex Numbers

Example: $(3 + 2i) + (4 + 3i) = 3 + 2i + 4 + 3i = 7 + 5i$

Example: $(6 + 4i) - (5 + 3i) = 6 + 4i - 5 - 3i = 1 + i$

IMAGINARY NUMBERS AND COMPLEX NUMBERS

Multiplying and Dividing Complex Numbers

1) $(x+yi)(a+bi) = x(a+bi) + yi(a+bi)$

$\quad = xa + (xb)i + (ya)i + (yb)$

$\quad = xa + (xb)i + (ya)i + (yb)(-1)$

$\quad = (xa-yb) + (xb+ya)i$

2) $(x+yi)(x-yi) =$

$\quad = x^2 - xyi + xyi - y^2i^2 = x^2 - y^2(-1)$

$\quad = x^2 + y^2$

Example: $(2 + i)(4 - i) = ?$

Solution:

$2 \times 4 - 2i + 4i - i^2 = 8 + 2i - (-1) = 8 + 2i + 1 = 9 + 2i$

3) When $b \neq 0$ and $c \neq 0$, always multiply the numerator and denominator by the opposite of the denominator.

$$\frac{a+bi}{c+di} = \frac{a+bi}{c+di}\left(\frac{c-di}{c-di}\right) = \frac{(ac+bd)+(bc-ad)i}{c^2+d^2}$$

Example: $\dfrac{2+3i}{2+i} = ?$

Solution:

$$\frac{2+3i}{2+i} = \frac{(2+3i)}{(2+i)} \times \left(\frac{2-i}{2-i}\right) = \frac{4-2i+6i-3i^2}{4-2i+2i-i^2} = \frac{4+4i+3}{4+1} = \frac{7+4i}{5}$$

IMAGINARY NUMBERS AND COMPLEX NUMBERS EXERCISES

Simplify each of the following.

1) $i^{15} + i^{21} + i^{48}$

2) $i^{20} - i^{26}$

3) $i^{2020} + i^{2021} + i^{2022}$

4) $i^2 - i^{12} + i^{18}$

5) $\dfrac{i^{150}}{i^{19}}$

6) $i^{2020} \times i^{2021}$

7) $\dfrac{3}{i}$

8) $(i - 2)(i + 2)$

9) $\dfrac{3}{2i} - \dfrac{5}{6i}$

10) $\sqrt{-625}$

11) $\sqrt{-36}$

12) $(i - 3)(i + 5)$

13) $\dfrac{3 - 2i}{5 + 2i}$

14) $\dfrac{1 - i}{1 + i}$

15) $\dfrac{5 - 2i}{3i}$

16) $\dfrac{-1 - 2i}{1 - 3i}$

17) $\dfrac{5}{-i}$

18) $\left(\dfrac{1 + i}{1 - i}\right)^{2018}$

SOLVING QUADRATIC EQUATIONS BY FACTORING

Foil Method:

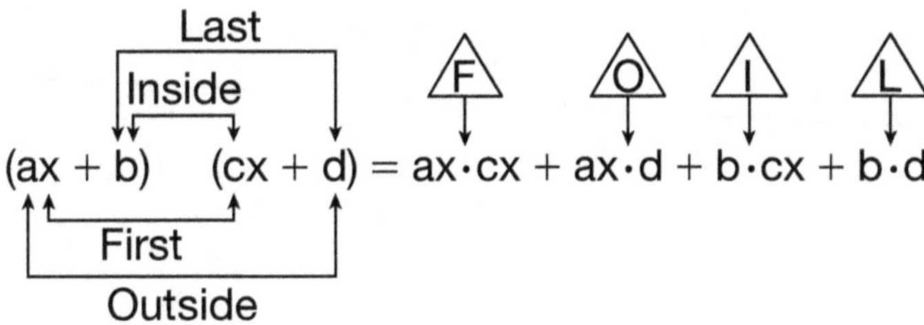

Examples:

1. $x^2 + 6x + 5 = (x + 5)(x + 1)$

2. $x^2 - 7x + 12 = (x - 4)(x - 3)$

3. $x^2 - 25 = (x - 5)(x + 5)$

Special Cases:

- $a^2 - b^2 = (a - b)(a + b)$

- $a^2 + b^2 = (a + b)^2 - 2ab$

- $a^3 - b^3 = (a - b)(a^2 + 2ab - b^2)$

- $a^3 - b^3 = (a - b)^3 + 3ab(a - b)$

- $a^3 + b^3 = (a + b)(a^2 - 2ab + b^2)$

- $a^3 + b^3 = (a + b)^3 - 3ab(a + b)$

- $\dfrac{a^2}{b^2} + \dfrac{b^2}{a^2} = \left(\dfrac{a}{b} + \dfrac{b}{a}\right)^2 - 2$

SOLVING QUADRATIC EQUATIONS BY FACTORING EXERCISES

Factor each of the following equations.

1) $x^2 - 49$

2) $x^2 - 100$

3) $x^2 - 8x + 15$

4) $x^2 - x + 42$

5) $6x^2 - 21x - 12$

6) $x^2 - 2$

7) $\dfrac{x^2}{y^2} + \dfrac{y^2}{x^2}$

8) $x^2 + 2xy + y^2$

9) $x^2 - 6x - 16$

10) $x^2 - 9x + 8$

11) $7x^2 + 55x - 72$

12) $2x^2 - 7x - 15$

13) $25x^2 + 5x - 2$

14) $x^2 - x - 3$

15) For what values of k is the following expression factorable?

$x^2 + kx + 3$

SOLVING QUADRATIC EQUATIONS BY QUADRATIC FORMULA AND COMPLETE SQUARE

Quadratic Equation: $ax^2 + bx + c$

Quadratic Formula: $x = \dfrac{-b \pm \sqrt{b^2 - 4ac}}{2a}$

Discriminant: $b^2 - 4ac$

- If $b^2 - 4ac < 0$, there are no real roots.

- If $b^2 - 4ac = 0$, there is 1 real root.

- If $b^2 - 4ac > 0$, there are 2 real roots.

Complete Square Method:

- Step 1: Write the equation in the form $ax^2 + bx = c$

- Step 2: Square half the coefficient of x, and add the result to both sides of the equation

- Step 3: Complete the square

- Step 4: Factor and simplify

- Step 5: Take the square root of both sides and solve equation

Example: $x^2 + 8x + 12 = 0$

- Step 1: $x^2 + 8x = -12$

- Step 2: $\dfrac{8}{2} = 4^2 = 16$

- Step 3: $x^2 + 8x + 16 = -12 + 16$

- Step 4: $(x+4)^2 = 4$

- Step 5: $(\sqrt{x+4})^2 = (\sqrt{4})^2$, then $x + 4 = \pm 2$, $x = -4 \pm 2$

  $x = -4 + 2$ or $x = -4 - 2$

  $x = -2$ or $x = -6$

SOLVING QUADRATIC EQUATIONS BY QUADRATIC FORMULA AND COMPLETE SQUARE EXERCISES

Solve each of following quadratic equations by using the quadratic formula.

$$x = \frac{-b \pm \sqrt{b^2 - 4ac}}{2a}$$

1) $x^2 + 8x + 12 = 0$

2) $x^2 + 2x + 1 = 0$

3) $4x^2 + 5x + 1 = 0$

4) $x^2 - 8x + 12 = 0$

5) $x^2 - x - 1 = 0$

6) $x^2 - 4x + 3 = 0$

7) $x^2 - 6x + 1 = 0$

8) $3x^2 + 4x - 3 = 0$

Solve each of the following equations by using the complete square method.

9) $x^2 + 4x + 3 = 0$

10) $x^2 + 6x - 16 = 0$

11) $x^2 + 12x + 8 = 0$

12) $x^2 - 6x - 20 = 0$

13) $2x^2 + 8x - 10 = 0$

14) $3x^2 + 6x - 18 = 0$

15) $5x^2 + 20x - 40 = 0$

16) $x^2 + 4x + 3 = 0$

INTEREST, MIXTURE, AND WORK PROBLEM SOLVING

Growth

$$A = P(1 + r)t$$

Decay

$$A = P(1 - r)t$$

A = Principal Amount

r = rate (as a decimal)

t = time

Work Problem Formula;

$$\frac{1}{x} + \frac{1}{y} = \frac{1}{t}$$

x = is the amount of time taken by first person to complete a job.

y = is the amount of time taken by first person to complete a job.

t = is the time taken if both do the together.

Example: Vera can paint a house in 12 hours and Nora can paint a house in 18 hours. If they worked together how long it take to paint a house?

Solution:

$$\frac{1}{V} + \frac{1}{N} = \frac{1}{t}$$

$$= \frac{1}{12} + \frac{1}{18} = \frac{1}{t}$$

$$= \frac{1}{12}\frac{3}{3} + \frac{1}{18}\frac{2}{2} = \frac{1}{t}$$

$$= \frac{3}{36} + \frac{2}{36} = \frac{1}{t},$$

$$\frac{5}{36} = \frac{1}{t}, \text{ then } t = \frac{36}{5} \text{ hours or } t = 7 . 2 \text{ hours}$$

INTEREST, MIXTURE, AND WORK PROBLEM SOLVING EXERCISES

1) Find the bank account balance if the account starts with $360, has an annual rate of 6%, and the money is left in the account for 5 years. (Round your answer to the nearest who¬le)

2) Melisa deposits $200 in her saving account with a rate of 6%. The interests compounded yearly. How much money will Melisa have after 8 years? (Round your answer to the nearest whole)

3) Jenny buys a car for $28,000. The value of the car decreases by 6% each year. Estimate the value after 5 years. (Round your answer nearest to whole)

4) Leyla has $358.60 in her checking account. How much does she have in her checking account after she makes a deposit of $140.20 and a withdrawal of $178.60?

5) An investment of $600 increases at a rate of 4% per year. Find the value of investment after 12 years.

6) How many liters of 80% pure water must be added to 40 liters of 30% pure water to produce 60% pure water?

7) The athlete moves from A to B and the dis¬tance is 41 of the total. When the athlete arrives at C, the distance equals 31 of the total. When the athlete arrives at B, the time is 8:50AM. When the athlete arrives at C, the time is 9:00AM. What time is it when the athlete arrives to D?

8) A new copy machine can print 120 pages per hour, and an older copy machine can print 80 pages per hour. How many minu¬tes will two copy machines working toget her, take to copy a total of 360 pages?

9) Vera can clean a house in 3 hours and Nora can clean a house in 4 hours. If they worked together how long it take to clean a house?

10) David can move the lawn in 1.5 hours and Tony can move the lawn in 3 hours. How long it will take for them to move the lawn together?

ARITHMETIC SEQUENCE AND GEOMETRIC SEQUENCE

Arithmetic Sequence: An arithmetic sequence is the difference between two consecutive terms.

$a_n = a_1 + (n - 1)d$

$a_n = n^{th}$ term $\qquad a_1 = $ first term $\qquad n = $ number of items $\qquad d = $ common difference

Example: Find the 45th terms in the arithmetic sequence 1, 5, 9, 13, 17, ...

Solution:

First term $a_1 = 1$

Common difference $d = 4$

Number of items $n = 45$

$a_n = a_1 + (n - 1)d$

$a_{45} = 1 + (45 - 1)4$

$a_{45} = 1 + 176$

$a_{45} = 177$

Geometric Sequence: A geometric sequence is the ratio between two consecutive terms.

$a_n = a_1(r)^{n - 1}$

$a_n = n^{th}$ term of the sequence $\qquad a_1 = $ first term $\qquad r = $ common ratio n: number of items

Example: If the first term of geometric sequence is 15 and common ratio is 6, then find 3rd term of the sequence.

Solution:

$a_n = n^{th}$ term $\qquad a_1 = 15 \qquad r = 5 \qquad n = 3$

$a_n = a_1(r)^{n - 1}$

$a_n = 15(6)^{n - 1}, \qquad a_3 = 15(6)^{3 - 1}, \qquad a_3 = 15(6)^2, \qquad a_3 = 15 \cdot 36, \quad a_3 = 540$

1) Find the 25th terms in the arithmetic sequence 4, 8, 12, 16, 20, …

2) Find the 30th terms in the arithmetic sequence –3, –10, –17, –24, …

3) If one term of arithmetic sequence $a_{48} = 120$ and the common difference d = 12. Find the first term.

4) If one term of arithmetic sequence $a_{18} = 96$ and the common difference d = 8. Find the first term.

5) If one term of arithmetic sequence $a_{10} = 48$ and $a_{25} = 140$. Find the first term.

6) If the first term of geometric sequence is 24 and a common ratio are 8. Find nth term formula of the sequence.

7) Determinate the nth formula of the geometric sequence below.

80, 40, 20, …

8) If the first term of geometric sequence is 95, and a common ratio are 2. Find 5th terms of the sequence.

9) Find the 7th term of the geometric sequence below.

6, 12, 24, 48,…

ARITHMETIC SERIES AND GEOMETRIC SERIES

Arithmetic Series:

$$S_n = \frac{n}{2}(a_1 + a_n)$$

a_1 = First term.

a_n = Last term.

n = Number of terms.

Example: Find the sum of the first 40 terms of the arithmetic series If $a_1 = 8$ and $a_{40} = 80$.

Solution:

$a_1 = 8$.

$a_n = a_{40} = 80$

n = 40

$$S_{40} = \frac{40}{2}(8 + 80) = 40 \cdot 88$$

$$S_{40} = 3320$$

Geometric Series:

S_n = sum of the first term.　　　　n = number of terms.　　　　r = common ratio

$$S_n = \frac{n(r^n - 1)}{r - 1} \ (r \neq 1).$$

Example: Find the sum of the first 5 terms of the geometric series if $a_1 = 15$ and $r = 3$.

Solution:

$a_1 = 15$, r=3, and n=5

$$S_n = \frac{n(r^n - 1)}{r - 1}$$

$$S_5 = \frac{5(3^5 - 1)}{3 - 1} = \frac{5(242)}{2} = 605$$

ARITHMETIC SERIES AND GEOMETRIC SERIES EXERCISES

1) Find the sum of the first 15 terms of the arithmetic series If $a_1 = 5$ and $a_{40} = 60$.

2) In the arithmetic series $7+ 14+ 21+ 28+ \ldots$ find the sum of first 30 terms.

3) Find the sum of the multiples of 5 between 10 and 30.

4) Find the sum of the first 4 terms of the geometric series if $a_1 = 10$ and $r = 2$.

5) Find the sum of the infinite geometric series $5 + 10 + 15 + 20 + 25 + \ldots$

6) Find a_1 if $s_n = 50$, $r = 4$, and $n = 3$

7) If $a_n = 2n - 1$, then what is the 4th term of this sequence?

8) What is the sum of the even integers between 35 to 75?

9) What is the missing value of x in the following sequence?

$$1, \frac{1}{2}, 0, -\frac{1}{2}, -1, x$$

The equation of a circle with its center at (0, 0) and with a radius r is:

$x^2 + y^2 = r^2$

Example: graph $x^2 + y^2 = 25$.

Solution:

$x^2 + y^2 = r^2$, From equation $r^2 = 25$, then r = 5

If r = 5 and center point is (0,0). Graph of equation:

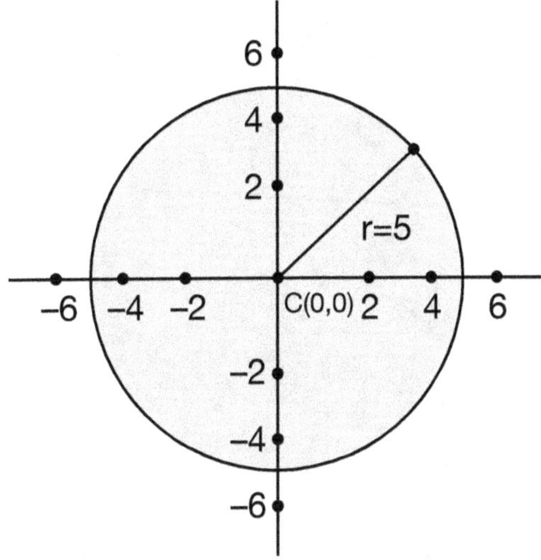

If the equation of a circle with its center at (h, k) and with a radius r is:

$(x - h)^2 + (y - k)^2 = r^2$

Example: graph the equation $(x - 4)^2 + (y - 5)^2 = 36$.

Solution:

$(x - h)^2 + (y - k)^2 = r^2$ $(x - 4)^2 + (y - 5)^2 = 36$.

$r^2 = 36$, then r=6. If r = 6 and center point is (4,5).

Graph of equation:

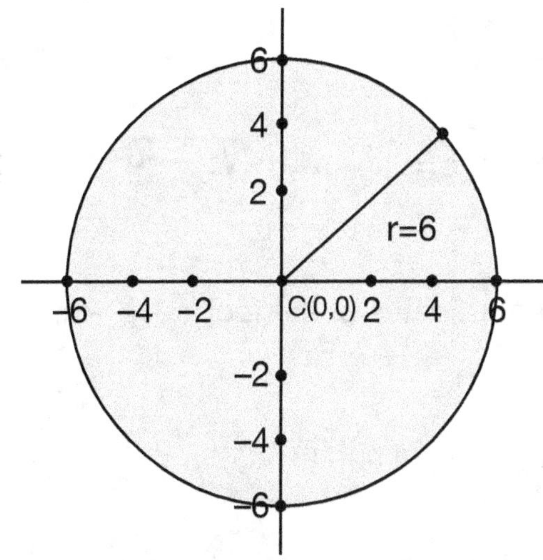

Horizontal major axis of ellipse:

$$\frac{(x-h)^2}{a^2} + \frac{(y-k)^2}{b^2} = 1$$

Vertical major axis of ellipse:

$$\frac{(x-h)^2}{b^2} + \frac{(y-k)^2}{a^2} = 1$$

Vertices: $(h + a, k)$ and $(h - a, k)$

Distance: $c = \sqrt{a^2 - b^2}$

Coordinates of the foci: $(h - c, k)$ and $(h + c, k)$

Example: Find the foci of $\frac{(x)^2}{144} + \frac{(y)^2}{81} = 1$

Solution:

We know the equation $\frac{(x-h)^2}{a^2} + \frac{(y-k)^2}{b^2} = 1$, the center is $(h, k) = (0,0)$ and $a^2 = 144$, $a = 12$ and $b^2 = 81, b = 9$.

Distance: $c = \sqrt{a^2 - b^2}, c = \sqrt{12^2 - 9^2}\ \sqrt{144 - 81} = \sqrt{63} = 3\sqrt{7}$

The coordinates of the foci are $(h - c, k)$ and $(h + c, k)$ then $(0 - 3\sqrt{7}, 0)$ and $(0 + 3\sqrt{7}, 0)$.

Foci: $(-3\sqrt{7}, 0)$ and $(3\sqrt{7}, 0)$.

PARABOLA

Equation of a Parabola:

- When it's open up or down;

  $(x - h)^2 = 4p(y - k)$ Where $p \neq 0$ with Vertex point (h, k)

  Directrix $y = k - p$ and Focus (h, k + p)

- When its open right or left;

  $(y + k)^2 = 4p(x - h)$ Where $p \neq 0$ with Vertex point (h, k),

  Directrix $x = h - p$ and Focus (h + p ,k)

- Quadratic equation form;

  $y = ax^2 + bx + c$

- Axis of symmetry is $x = -\dfrac{b}{2a}$

Example: Write the equation of parabola $x^2 + 12x - 4y + 20 = 0$ in standard form and find vertex point.

Solution:

$x^2 + 12x - 4y + 20 = 0$

$x^2 + 12x + 20 = 4y$

$(x + 6)^2 - 36 + 20 = 4y$

$(x + 6)^2 = 4y + 16$

Standard form: $(x + 6)^2 = 4(y + 4)$

Vertex point: h, k = (−6, −4)

1) From the equation of circle below find the radius.

$$x^2 + y^2 + 2x - 6y = 20$$

2) From the equation of circle below find the center point.

$$x^2 + 10x + y^2 + 4y - 18 = 0$$

3) Find the standard form for the equation of the circle centered at (3, 6) and a radius of $2\sqrt{5}$.

4) Find the foci of: $\dfrac{(x)^2}{25} + \dfrac{(y)^2}{9} = 1$

5) Find the vertices of: $\dfrac{(x)^2}{36} + \dfrac{(y)^2}{16} = 1$

6) Find the foci of: $\dfrac{(x-2)^2}{25} + \dfrac{(y+3)^2}{4} = 1$

7) Write the equation of parabola $x^2 + 8x + y^2 - 4y + 36 = 0$ in standard form and find vertex point.

8) What is the vertex point of the following parabola?

$$y = x^2 + 6$$

9) Find the focus for the parabola.

$$(x-3)^2 = 4p(y - 5)$$

10) Find the directrix for the parabola.

$$(y + 2)^2 = 4p(x + 7)$$

11) Write the equation of following parabola.

Vertex=(0, 0) and focus = (1, 4)

LOGARITHMS AND OPERATIONS WITH LOGARITHMS

$y = \log_a x \Leftrightarrow x = a^y$

$\log_a x^b = b \log_a x$

$\log_a a = 1$

$\log_a 1 = 0$

$\log_a 0 = \text{Undifined}$

Example: Evaluate $\log_2 32$.

Solution:

$\log_a x^b = b \log_a x$

$\log_2 32 = \log_2 2^5 = 5 \overset{1}{\cancel{\log_2 2}} = 5$

Product Property

$\log_a x \cdot y = \log_a x + \log_a y$

Example: Simplify $\log_4 (3 \cdot 5)$.

Solution:

$\log_a x \cdot y = \log_a x + \log_a y$

$\log_4 (3 \cdot 5) = \log_4 3 + \log_4 5$

Quotient Property

$\log_a \left(\dfrac{x}{y} \right) = \log_a x - \log_a y$

Example: Simplify $\log_3 \left(\dfrac{5}{7} \right)$.

Solution:

$\log_3 \left(\dfrac{5}{7} \right) = \log_3 5 - \log_3 7$

Power Property

$\log_a (x^n) = n \log a^x$

Example: Simplify $\log_7 (2^3)$.

Solution:

$\log_a (x^n) = n \log_a x$

$\log_7 (2^3) = 3 \log_7 2$

The change of base property

$\log_a x = \dfrac{\log_k x}{\log_k a}$

Example: Simplify $\log_3 7$.

Solution:

$\log_a x = \dfrac{\log_k x}{\log_k a}$

$\log_3 7 = \dfrac{\log_{10} 7}{\log_{10} 3}$

Natural Logarithms

$\ln 1 = 0$

$\ln e = 1$

$\ln e^x = x$

Example: If $\ln x^3 = 2 + \ln y^5$, then express in terms of x.

Solution: $\ln x^3 - \ln y^5 = 2$

$\ln\left(\dfrac{x^3}{y^5}\right) = 2,$

$e^2 = \dfrac{x^3}{y^5}, \quad x^3 = e^2 \cdot y^5, \text{ then } x = \sqrt[3]{e^2 y^5}$

LOGARITHMS AND OPERATIONS WITH LOGARITHMS
EXERCISES

1) Evaluate the expression $\log_4 32$.

2) Solve equation $\log_{16}(x) = \dfrac{3}{4}$.

3) Simplify $\log_3 40 - \log_3 5$

4) If $\log_{10} 40 - \log_{10} x + 1 = 1$, then find x.

5) Find the value of x in the following equation.

$$\log_2 x - 2 + \log_2 x + 3 = \log_2 6$$

6) What is the value of x in $\log_2 3x - 1 = 5$.

7) Solve $\dfrac{1}{\log_4 8} + \dfrac{1}{\log_4 8}$.

8) Evaluate $\dfrac{3\log_5 27}{4\log_5 3}$.

9) Evaluate $\log_{25} x = \dfrac{3}{2}$.

MATRICES AND OPERATION WITH MATRICES

Row Matrix: A matrix having a single row with horizontal arrays is called a row matrix.

Example:

(a b c)

Column matrix: A matrix having a single column with vertical arrays is called a column matrix.

Example:

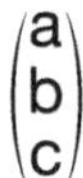

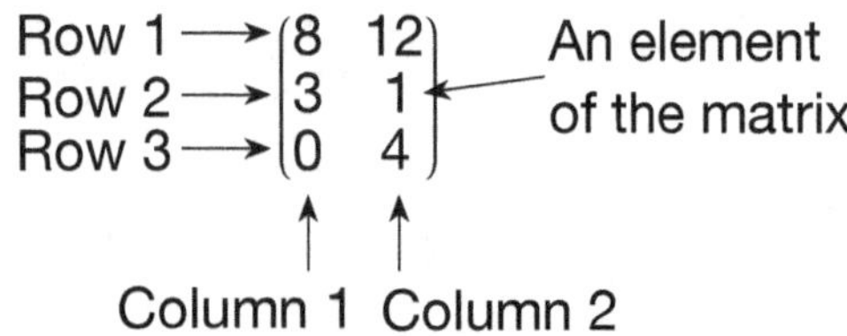

dimension of this matrix is 3×2

Adding Matrices

$$A = \begin{bmatrix} a & b \\ c & d \end{bmatrix} \quad B = \begin{bmatrix} e & f \\ g & h \end{bmatrix}$$

$$A + B = \begin{bmatrix} a & b \\ c & d \end{bmatrix} + \begin{bmatrix} e & f \\ g & h \end{bmatrix} = \begin{bmatrix} a+e & b+f \\ c+g & d+h \end{bmatrix}$$

Subtracting Matrices

$$A = \begin{bmatrix} a & b \\ c & d \end{bmatrix} \quad B = \begin{bmatrix} e & f \\ g & h \end{bmatrix}$$

$$A - B = \begin{bmatrix} a & b \\ c & d \end{bmatrix} - \begin{bmatrix} e & f \\ g & h \end{bmatrix} = \begin{bmatrix} a-e & b-f \\ c-g & d-h \end{bmatrix}$$

Multiplying Matrices

$$A = \begin{bmatrix} a & b \\ c & d \end{bmatrix} \quad B = \begin{bmatrix} k & l \\ m & n \end{bmatrix}$$

$$A \cdot B = \begin{bmatrix} a & b \\ c & d \end{bmatrix} - \begin{bmatrix} k & l \\ m & n \end{bmatrix} = \begin{bmatrix} ak+bm & a \cdot i + b \cdot n \\ ck+dm & c \cdot i + d \cdot n \end{bmatrix}$$

Matrix Determinant

Symbol of determinant $|A|$

$$|A| = \begin{vmatrix} a & b \\ c & d \end{vmatrix} = a \cdot d - b \cdot c$$

Example:

Let $K = [2 \ 4]$ and $L = \begin{bmatrix} -3 & 4 \\ 2 & -1 \end{bmatrix}$, then find $K \cdot L$

Solution:

$$K \cdot L = [2 \ 4] \cdot \begin{bmatrix} -3 & 4 \\ 2 & -1 \end{bmatrix} = [2 \cdot (-3) + 4 \cdot 2 \quad 2 \cdot 4 + 4 \cdot (-1)]$$

$$= [2 \ 4]$$

1) If $3\begin{bmatrix} x \\ y \end{bmatrix} = 4\begin{bmatrix} 6 \\ 9 \end{bmatrix}$, then what is the value of x + y?

2) If matrix $y = \begin{bmatrix} 3 & 6 \\ 4 & 7 \end{bmatrix}$ and matrix $x - y = \begin{bmatrix} 1 & 3 \\ 2 & 4 \end{bmatrix}$, then find matrix x?

3) If matrix $A = \begin{bmatrix} a & b \\ c & d \end{bmatrix}$ and matrix $B = \begin{bmatrix} e & f \\ g & h \end{bmatrix}$, then find matrix A + B?

4) In the equation below, find x in terms of y?

$$\begin{bmatrix} x & y \\ 2 & 4 \end{bmatrix} = 10$$

5) If $\begin{bmatrix} 1 & 3 \\ -2 & 4 \end{bmatrix}\begin{bmatrix} x \\ y \end{bmatrix} = \begin{bmatrix} 1 \\ 8 \end{bmatrix}$, then find x + y?

6) If Matrix $x = \begin{bmatrix} 1 & 3 \\ -2 & 4 \end{bmatrix}$ and matrix $y = \begin{bmatrix} 1 & 3 \\ -2 & 4 \end{bmatrix}$, then find x·y?

FINAL TEST

1. If $a = \sqrt{5}$ and $b = \sqrt{3}$, then find $\dfrac{a}{b} - \dfrac{b}{a}$.

A) $\dfrac{\sqrt{15}}{15}$

B) $\dfrac{2\sqrt{15}}{15}$

C) $\dfrac{\sqrt{15}}{5}$

D) $\dfrac{4\sqrt{15}}{15}$

2. Simplify $\dfrac{\sqrt{x}}{2+\sqrt{x}}$?

A) $\dfrac{2\sqrt{x}-x}{4-x}$

B) $\dfrac{\sqrt{x}-x}{4-x}$

C) $\dfrac{\sqrt{x}-x}{x}$

D) $\dfrac{2}{4-x}$

3. If $\dfrac{x-2}{y-3} = \dfrac{2}{3}$, then find y in terms of x?

A) $3x$

B) $5x$

C) $\dfrac{1}{2}x$

D) $\dfrac{3}{2}x$

4. The interior angles of a triangle ratio are 8 : 6 : 4. What is the measure of largest angle?

A) 30°

B) 40°

C) 60°

D) 80°

5. If y varies inversely as x and x = 10 when y = 18, find y when x = 36.

A) 1

B) 3

C) 5

D) 7

6. In a class of 20 students, 45% received B's and 30% received A's and rest of class received C's in Science Test. How many students received A?

A) 3

B) 5

C) 6

D) 8

7. Write 0.015 as a percent.

A) 0.15%

B) 1.5%

C) 15%

D) 150%

9. The school basketball team played 55 games and lose11 of them. What percent of the games did they lose?

A) 20%

B) 40%

C) 60%

D) 80%

8. If $\dfrac{x}{\left(\frac{2}{5}\right)} = \dfrac{\left(\frac{1}{8}\right)}{\left(\frac{3}{2}\right)}$ Find x.

A) $\dfrac{1}{15}$

B) $\dfrac{1}{30}$

C) $\dfrac{1}{45}$

D) $\dfrac{2}{15}$

10. If x is a negative number, and $x^2 + x - 12 = 0$, what is the value of x?

A) −2

B) −3

C) −4

D) −6

11. If $\left|\dfrac{x}{4} - 6\right| < 7$ what is a possible value of x in the above inequality?

A) −4

B) −10

C) −15

D) 25

12. If $3x - 10 \leq 8$, then what is the largest possible value of $3x + 1$?

A) 13

B) 14

C) 15

D) 19

13. If $x > 0$, then $\dfrac{|x| + |-2x|}{|-3x|}$ is equal to which of following.

A) −1

B) 0

C) 1

D) 2

14. If x is the greatest prime factor of 26 and y is the greatest prime factor of 55, what is the value of x + y?

A) 8

B) 12

C) 24

D) 30

15. The population of a city increased from 270 thousand to 360 thousand in 2017 to 2018. Find the percent of increase.

A) $13.\overline{3}\%$

B) $23.\overline{3}\%$

C) $33.\overline{3}\%$

D) $43.\overline{3}\%$

American Math Academy

16.

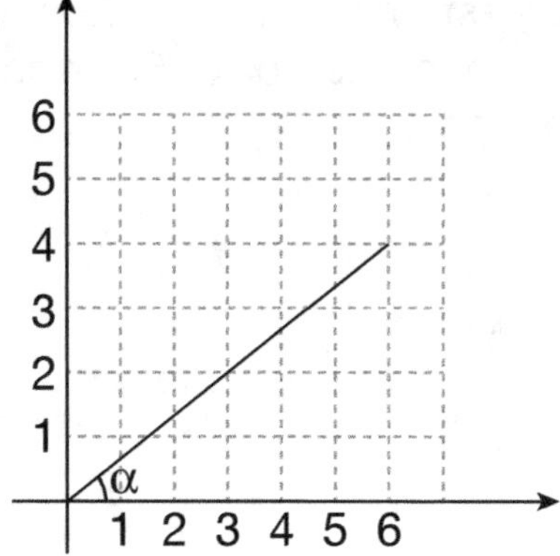

What is the equation of the function?

A) $y = 2x$

B) $y = \dfrac{2}{3}x$

C) $y = x + 7$

D) $y = 2x + 3$

17.
$$ax^3 + bx^2 + cx + d = 0$$

In the equation above, a,b,c and d are constants. If the equation roots are –2,4 and –7, which of the following is a factor of $ax^3 + bx^2 + cx + d$?

A) $x - 2$

B) $x + 4$

C) $x + 7$

D) $x - 7$

18. $A\{0, 1, 2, 3, 4, 5, 6, 7, 8, 9\}$

What is the probability of selecting an even number in set A?

A) $\dfrac{1}{2}$

B) $\dfrac{3}{4}$

C) $\dfrac{5}{6}$

D) $\dfrac{2}{3}$

19. If $3y - \dfrac{x}{4} = 10$, then which of the following is equal to $\dfrac{x}{2}$?

A) $6y - 20$

B) $3y - 10$

C) $y - 20$

D) $6y + 20$

20. If $x^{\frac{1}{3}} = 64$, then find $\dfrac{x}{2}$.

A) 2^{15}

B) 2^{17}

C) 2^{18}

D) 2^{20}

21. If x, y and z are real numbers and;

$$x^3 \cdot y^2 > 0$$

$$x^2 \cdot z > 0$$

$$y^3 \cdot z < 0$$

Which of the following must be true?

A) +, +, +

B) −, +, −

C) +, −, −

D) +, −, +

22. Simplify $\sqrt{27} - \sqrt{81} + \sqrt{243}$.

A) $3\sqrt{3} - 9$

B) $6\sqrt{3} - 9$

C) $9\sqrt{3} - 9$

D) $12\sqrt{3} - 9$

23. The price of a book has been discounted 20%. The sale price is $45. What is the original price?

A) $45.5

B) $51.5

C) $51.75

D) $56.25

24. Find y when x = 3, if y varies directly as x and y = 20 when x = 5.

A) 6

B) 9

C) 12

D) 15

25. If $\dfrac{a^2}{b^2} + \dfrac{b^2}{a^2} = 7$, then which of following could be $\dfrac{a}{b} + \dfrac{b}{a} = ?$

A) 1

B) 3

C) 2

D) −1

26. Evaluate $4 + (2 + 3) \cdot 5 - 8 + (5 + 3)^0$?

A) 15

B) 16

C) 17

D) 22

27. If $x > 0$ and $x - 3 = \sqrt{x-3}$, then which of following can be x?

A) 0

B) 1

C) 2

D) 4

28. Solve $\dfrac{2}{2-\sqrt{2}}$.

A) 4

B) $-\sqrt{2}$

C) $2 + \sqrt{2}$

D) $2 - \sqrt{2}$

29. Simplify $\dfrac{1}{x-1} + \dfrac{1}{x+1}$.

A) $x - 1$

B) $x + 1$

C) 1

D) $\dfrac{2x}{x^2-1}$

30. Simplify $\dfrac{x^2-8x+15}{x^2-9} \div \dfrac{x^2-4x-5}{x^2+3x}$.

A) $x - 1$

B) $\dfrac{x}{x+1}$

C) $x + 1$

D) $\dfrac{x+1}{x-1}$

31.

$$x + 3ky = 12$$
$$5x - 12y = 18$$

In the system of equations above, k is a constant. For what value of k will the system of equations have no solutions?

A) $-\dfrac{4}{5}$

B) $\dfrac{4}{5}$

C) $\dfrac{5}{4}$

D) $-\dfrac{5}{4}$

32. In the polynomial below, a is the constant. If the polynomial P(x) is divisible by x+2 then find the value of a.

$$P(x) = 2x^3 + ax^2 - x + 2$$

A) 2

B) 3

C) 5

D) 7

33.

$$i^{2019} + i^{2020} + i^{2021}$$

Which of the following is equivalent to the complex number shown above?

A) i

B) 1

C) 2i

D) 2

34. An investment of $600 increases at a rate of 4% per year. Find the value of investment after 12 years. (Round your answer to the nearest dollar).

A) $941

B) $951

C) $961

D) $992

35. If $P(x-2) = x^2 + 2x + 1$, then find $P(x+1)$?

A) $x^2 + 8x + 16$

B) $-x^2 + 8x + 16$

C) $x^2 - 8x + 16$

D) $x^2 + 8x - 16$

36. $A = 3x + 1 = 4y + 2 = 5z + 3$

From the above equations x, y, and z are positive integer and A is a two digit number. What is the smallest value of A?

A) 58

B) 68

C) 78

D) 88

37. Melisa makes an online purchase SAT book and 25% discount is applied to the book price, then 4% tax is added to this discounted price. Which of the following represents the amount Melisa pays for an item with a book price in d dollars?

A) 0.75d

B) 0.78d

C) 0.88d

D) 0.92d

38. $\triangle = x^2 - 2x + 1$

$\bigcirc = x^2 - 1$

From above, if $\bigcirc = \triangle$, then find x.

A) 1

B) 2

C) 3

D) 4

39.

What is the equation of the function?

A) $y = 2x$

B) $y = \dfrac{2}{3}x$

C) $y = x + 7$

D) $y = 2x + 3$

40. $\dfrac{1}{a} = \dfrac{1}{b} + \dfrac{1}{c}$ Find b in terms of a and c.

A) $b = \dfrac{ac}{c-a}$

B) $b = \dfrac{ac}{a-c}$

C) $b = ac$

D) $b = \dfrac{1}{c-a}$

41. If $\dfrac{x}{y} = \dfrac{a}{b} = \dfrac{2}{3}$ and $y^2 - b^2 = 27$, then what is the value of $x^2 - a^2$?

A) 12

B) 20

C) 24

D) 25

42 If $a = 1 - 5i$ and $b = 1 + 5i$, then which of the following is equal to $a \cdot b$?

A) 10

B) 15

C) 25

D) 26

43. The price of a book has been discounted 20%. The sale price is $60. What is the original price?

A) $25

B) $45

C) $65

D) $75

44. If $\cos x + \sin x = \sqrt{5} + 1$, then find $\cos x \cdot \sin x = ?$

A) $\dfrac{5 - 2\sqrt{5}}{2}$

B) $\dfrac{5 + 2\sqrt{5}}{2}$

C) $5 - \sqrt{5}$

D) $2 + \sqrt{5}$

45. If $A = \begin{bmatrix} a & b \\ c & d \end{bmatrix}$ and $B = \begin{bmatrix} e & f \\ g & h \end{bmatrix}$, then which of following is A - B?

A) $\begin{bmatrix} a & b \\ c & d \end{bmatrix}$

B) $\begin{bmatrix} e & f \\ g & h \end{bmatrix}$

C) $\begin{bmatrix} a & f \\ b & h \end{bmatrix}$

D) $\begin{bmatrix} a-e, & b-f \\ c-g, & d-h \end{bmatrix}$

46. If $x^3 = y$, then find $\log_y x^2$?

A) 1

B) 2

C) $\dfrac{1}{3}$

D) $\dfrac{2}{3}$

47. $\begin{bmatrix} a & 6 \\ b & 4 \end{bmatrix} = 18$

In the equation above what is the a in terms of b?

A) $\dfrac{9 + 3b}{2}$

B) $\dfrac{9 + 2b}{3}$

C) $\dfrac{2 + 3b}{9}$

D) $\dfrac{2 + 9b}{3}$

48. In the following figure P is the center of circle, find the area of the shaded region.

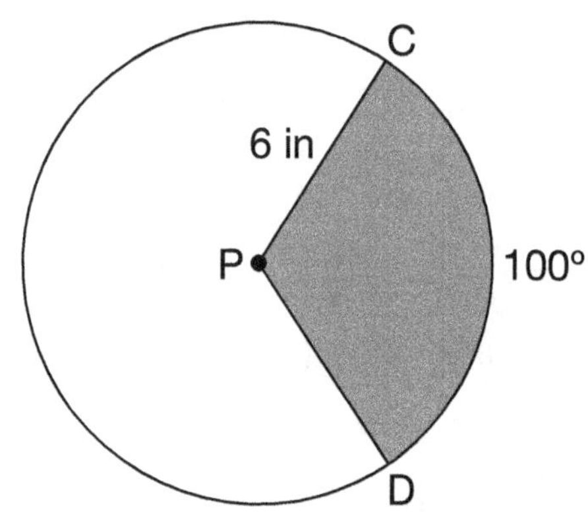

A) 5π

B) 10π

C) 20π

D) 30π

49. $\dfrac{a^{2m}}{a^{10}} = a^6$ and $a^{3n} = a^{30}$, then $m \cdot n$?

A) 50

B) 60

C) 70

D) 80

50. $\dfrac{1}{3}(6x - 9) + (x - 12) = ax + x + b$

What is the value of $a - b$?

A) 6

B) 9

C) 12

D) 17

51. Which of the following is equivalent to the expression below?

$(x^2y - 2x^2 + 6xy) - (xy^2 - 2x^2 + 4xy)$

A) $x^2y - xy^2 + 2xy$

B) $x^2y - xy^2 - 2xy$

C) $x^2y + xy^2 + 2xy$

D) $-x^2y - xy^2 - 2xy$

52. If $\dfrac{3}{4}x - \dfrac{1}{8}x = \dfrac{1}{12} + \dfrac{2}{3}$, what the value of x?

A) 6

B) 7

C) $\dfrac{6}{7}$

D) $\dfrac{6}{5}$

53. Which of the following is equivalent to the complex number $\dfrac{3+i}{2-i}$?

A) $1 - i$

B) 1

C) i

D) $1 + i$

SAMPLE TEST SOLUTIONS

1. Solution:

$\dfrac{x+2}{5} = \dfrac{x-3}{4}$ (cross – multiply)

$4(x + 2) = 5(x - 3)$

$4x + 8 = 5x - 15$

$8 + 15 = 5x - 4x$

$23 = x$

2. Solution:

$-3x - 18 + 20 \le 2(x - 8)$

$-3x + 2 \le 2x - 16$

$16 + 2 \le 2x + 3x$

$18 \le 5x$

$\dfrac{18}{5} \le x$

3. Solution:

$ab = \cancel{b} + c$

$ -b \quad \cancel{-b}$

$+ \overline{}$

$ab - b = c$

$\dfrac{b(\cancel{a-1})}{\cancel{a-1}} = \dfrac{c}{a-1} , b = \dfrac{c}{a-1}$

4. Solution:

$\left| \dfrac{1}{2}x \right| \le 9 , -9 \le \dfrac{1}{2}x \le 9$

$-18 \le x \le 18$

5. Solution:

$6x^2 - 2x - (2x - 5x^2)$

$= 6x^2 - 2x - 2x + 5x^2$

$= 11x^2 - 4x$

6. Solution:

$\dfrac{\cancel{2}5x^{-8}y^{10}}{\cancel{5}x^{10}y^{-5}} = \dfrac{5x^{-8}y^{10}}{x^{10}y^{-5}}$

$5x^{-8-10}y^{10-(-5)}$

$5x^{-18}y^{15}$

$= \dfrac{5y^{15}}{x^{18}}$

7. Solution:

$= (x^3 + 4x^2) - (3x + 12)$

$= x^2(x + 4) - 3(x + 4)$

$= (x + 4)[x^2 - 3]$

$= (x + 4) \cdot (x - \sqrt{3})(x + \sqrt{3})$

8. Solution:

$\dfrac{x^2 - 25}{x^3 - 125} = \dfrac{(x - 5)(x + 5)}{(x^3 - 5^3)}$

$= \dfrac{(\cancel{x - 5})(x + 5)}{(\cancel{x - 5})(x^2 + 5x + 25)} = \dfrac{x + 5}{x^2 + 5x + 25}$

SAMPLE TEST SOLUTIONS

9. Solution:

$$\frac{x+1}{x^2-1} - \frac{x-1}{x^2+1}$$

$$\frac{x+1}{(x-1)(x+1)} - \frac{x-1}{x^2+1}$$

$$\frac{1(x^2+1)}{(x-1)(x^2+1)} - \frac{(x-1)(x-1)}{(x^2+1)(x-1)}$$

$$\frac{x^2+1}{x^3-x^2+x-1} - \frac{x^2-x-x+1}{x^3-x^2+x-1}$$

$$\frac{x^2+1-x^2-x-x-1}{x^3-x^2+x-1}$$

$$\frac{2x}{x^3-x^2+x-1}$$

11. Solution:

$$\frac{x}{\dfrac{x\cdot 5}{3\cdot 5} - \dfrac{x\cdot 3}{5\cdot 3}} = \frac{x}{\dfrac{5x-3x}{15}}$$

$$\frac{x}{1}\cdot\frac{15}{2x} = \frac{15x}{2x} = \frac{15}{2}$$

13. Solution:

$$9^{2x-3} = 81^{3x-8}$$

$$(3^2)^{2x-3} = (3^4)^{3x-8}$$

$$\overset{1}{2}(2x-3) = \overset{2}{4}(3x-8)$$

$$2x-3 = 6x-16$$

$$\begin{array}{r} -2x \quad\quad -2x \\ \hline -3 = 4x-16 \end{array}$$

$$\begin{array}{r} 16 \quad\quad +16 \\ \hline \end{array}$$

$$13 = 4x, \text{ then } x = \frac{13}{4}$$

American Math Academy

10. Solution:

$$\frac{y}{y-1} \div \frac{y^2}{y^3-y}$$

$$\frac{y}{y-1} \cdot \frac{y(y^2-1)}{y^2}$$

$$\frac{y}{y-1} \cdot \frac{y(y-1)(y+1)}{y^2}$$

$$y+1$$

12. Solution:

$$g(x) = 3x-5$$

$$g(-3x) = 3(-3x)-5$$

$$= -9x-5$$

14. Solution:

$$\text{From } \frac{D}{180°} = \frac{R}{\pi}, \text{ then } \frac{D}{180°} = \frac{\dfrac{\pi}{5}}{\dfrac{\pi}{1}}$$

$$\frac{D}{180°} = \frac{1}{5} \text{ (cross–multiply)}$$

$$5D = 180$$

$$D = 36°$$

15. Solution:

$$\sqrt{\frac{64x^6y^8}{16x^1y^1}}$$

$$=\sqrt{\frac{64x^{6-1}y^{8-1}}{16}}$$

$$=\sqrt[2]{4x^5y^7}=\sqrt{2^2x^5y^7}$$

$$=2x^2y^3\sqrt{xy}$$

17. Solution:

$$\frac{\sqrt{x}+\sqrt{y}}{\sqrt{x}-\sqrt{y}}$$

$$=\frac{\sqrt{x}+\sqrt{y}}{\sqrt{x}-\sqrt{y}}\cdot\frac{\sqrt{x}+\sqrt{y}}{\sqrt{x}+\sqrt{y}}$$

$$=\frac{\sqrt{x^2}+\sqrt{xy}+\sqrt{xy}+\sqrt{y^2}}{\sqrt{x^2}-\sqrt{xy}+\sqrt{xy}-\sqrt{y^2}}$$

$$=\frac{x+2\sqrt{xy}+y}{x-y}$$

19. Solution:

$$\boxed{i^2=-1}$$

$$i^2-i^{12}+i^{18}$$

$$(i^2)^1-(i^2)^6+(i^2)^9$$

$$(-1)^1-(-1)^6+(-1)^9$$

$$=-1-1-1=-3$$

16. Solution:

$$\sqrt{48}-\sqrt{12}$$

$$=\sqrt{16{\times}3}-\sqrt{3{\times}4}$$

$$=4\sqrt{3}-2\sqrt{3}$$

$$=(4-2)\sqrt{3}$$

$$=2\sqrt{3}$$

18. Solution:

$$(x^{-10}y^5)^{\frac{-1}{5}}$$

$$=x^{\frac{10}{5}}y^{\frac{-5}{5}}$$

$$x^2y^{-1}=\frac{x^2}{y}$$

20. Solution:

	liters pure water	% water	total liters
80% water	x	.80	.80x
30% water	40	.30	.30(40)=12
60% water	x + 40	.60	.60(x+40)

From the last column, you get the equation

0.80x + 12 = 0.6(x + 40) Solve for x.

0.80x + 12 = 0.6x + 24

0.80x − 0.6x = 24 − 12

0.2x = 12

2x = 120

x = 60 liters.

21. Solution:

$a_n = a_1 + (n-1)d$

$a_{48} = a_1 + (48 - 1)12$

$120 = a_1 + (48 - 1)12$

$120 = a_1 + (47)12$

$120 = a_1 + 564$

$120 - 564 = a_1$, then $a_1 = -444$

22. Solution:

$r = \dfrac{10}{5} = 2,\ n = 5$

$S_5 = Sn = \dfrac{n(r^n - 1)}{r - 1}$

$S_5 = \dfrac{5(2^5 - 1)}{2 - 1} = \dfrac{5 \times 31}{1}$

$S_5 = 155$

23. Solution:

$x^2 + y^2 + 2x - 6y = 20$

$x^2 + 2x + y^2 - 6y = 20$

$(x + 1)^2 - 1 + (y - 3)^2 - 9 = 20$

$(x + 1)^2 + (y - 3)^2 = 20 + 10$

$(x + 1)^2 + (y - 3)^2 = 30$

$(x - h)^2 + (y - k)^2 = r^2$

$r^2 = 30$

$r = \sqrt{30}$

24. Solution:

$\log_2 3x - 1 = 5$

$3x - 1 = 2^5,\ 3x - 1 = 32$

$3x = 33$

$x = 11$

25. Solution:

$x \cdot y = \begin{bmatrix} 1 & 3 \\ -2 & 4 \end{bmatrix} \cdot \begin{bmatrix} 1 & 3 \\ -2 & 4 \end{bmatrix}$

$= \begin{bmatrix} 1 \cdot 1 + 3 \cdot (-2) & 1 \cdot 3 + 3 \cdot 4 \\ -2 \cdot 1 + 4 \cdot (-2) & (-2) \cdot 3 + 4 \cdot 4 \end{bmatrix}$

$= \begin{bmatrix} 1 - 6 & 3 + 12 \\ -2 - 8 & -6 + 16 \end{bmatrix}$

$= \begin{bmatrix} -5 & 15 \\ -10 & -10 \end{bmatrix}$

SOLVING LINEAR EQUATIONS
EXERCISES SOLUTIONS

1. Solution:

$$5x - 15 = 25$$
$$+15 \quad 15$$
$$\overline{}$$
$$5x = 40$$
$$x = 8$$

2. Solution:

$$3x - 30 = 45$$
$$+30 \quad 30$$
$$\overline{}$$
$$3x = 75$$
$$x = 25$$

3. Solution:

$$x - 10 = -25$$
$$+10 \quad +10$$
$$\overline{}$$
$$x = -15$$

4. Solution:

$$x + 30 = -15$$
$$-30 \quad -30$$
$$\overline{}$$
$$x = -45$$

5. Solution:

$$-2x - 5 = -35$$
$$+5 \quad +5$$
$$\overline{}$$
$$-2x = -30$$
$$x = 15$$

6. Solution:

$$-3x + 3 = -18$$
$$-3 \quad -3$$
$$\overline{}$$
$$-3x = -21$$
$$x = 7$$

7. Solution:

$$\frac{x}{2} + 3 = -4$$
$$-3 \quad -3$$
$$\overline{}$$
$$\frac{x}{2} = -7$$
$$x = -14$$

8. Solution:

$$-\frac{x}{4} - 5 = -9$$
$$+5 \quad +5$$
$$\overline{}$$
$$-\frac{x}{4} = -4$$
$$-x = -16, \; x = 16$$

9. Solution:

$\dfrac{x}{5} - \dfrac{3}{4} = -1$ (Multiply all equation by 20)

$20(\dfrac{x}{5} - \dfrac{3}{4} = -1)$

$\dfrac{20x}{5} - \dfrac{60}{4} = -20$

$4x - 15 = -20$

$ + 15 \quad +15$

$+$ ——————————

$4x = -5, \quad x = \dfrac{-5}{4}$

10. Solution:

$\dfrac{-x}{2} - \dfrac{7}{3} = -11$ (Multiply all equation by 6)

$6\left(\dfrac{-x}{2} - \dfrac{7}{3} = -11\right)$

$\dfrac{-6x}{2} - \dfrac{42}{3} = -66$

$-3x - 14 = -66$

$+14 \quad +14$

$+$ ——————————

$-3x = -52$

$x = \dfrac{52}{3}$

11. Solution:

$5x - 2(x + 5) = 4(2x + 6) + 8$

$5x - 2x - 10 = 8x + 24 + 8$

$3x - 10 = 8x + 32$

$-3x -3x$

$+$ ——————————

$-10 = 5x + 32$

$-32 -32$

$+$ ——————————

$-42 = 5x$

$\dfrac{-42}{5} = x$

12. Solution:

$6x - 11 = 5x + 19$

$-5x -5x$

——————————

$x - 11 = +19$

$+ \quad 11 = +11$

——————————

$x = 30$

13. Solution:

$\dfrac{x + 2}{5} = \dfrac{x - 3}{4}$ (cross – multiply)

$4(x + 2) = 5(x - 3)$

$4x + 8 = 5x - 15$

$23 = x$

14. Solution:

$\dfrac{x + 3}{2x} = \dfrac{2}{3}$ (cross – multiply)

$3x + 9 = 4x$

$9 = x$

15. Solution:

$$-3(2x + 7) = 4(5x + 10)$$

$$-6x - 21 = 20x + 40$$

$$+6x \qquad\qquad +6x$$

$$\overline{}$$

$$-21 = 26x + 40$$

$$-40 \qquad\qquad -40$$

$$+\overline{}$$

$$-61 = 26x, \quad x = \frac{-61}{26}$$

16. Solution:

$$\frac{1}{2}(2x + 10) = -4(5x + 10)$$

$$x + 5 = -20x - 40$$

$$-x \qquad\qquad -x$$

$$+\overline{}$$

$$5 = -21x - 40$$

$$40 \qquad\qquad +40$$

$$+\overline{}$$

$$45 = -21x, \quad x = \frac{-45}{21} = \frac{-15}{7}$$

17. Solution:

$$\frac{3x - 7}{5} - \frac{x}{3} = \frac{2}{1} - \frac{x - 3}{4}$$

$$\frac{(3x - 7)3}{5 \cdot 3} - \frac{x(5)}{3(5)} = \frac{2 \cdot 4}{1 \cdot 4} - \frac{x - 3}{4}$$

$$\frac{9x - 21 - 5x}{15} = \frac{8 - x + 3}{4}$$

$$\frac{4x - 21}{15} = \frac{11 - x}{4} \quad \text{(cross – multiply)}$$

$$16x - 84 = 165 - 15x, \quad 31x = 249$$

$$x = \frac{249}{31}$$

18. Solution:

$$\frac{1}{x} = \frac{3x + 10}{4x}$$

$$4 \cdot 1 = 1(3x + 10)$$

$$4 = 3x + 10$$

$$-10 \qquad\qquad -10$$

$$+\overline{}$$

$$-6 = 3x$$

$$-2 = x$$

1. Solution:

$$x - 15 < 25$$
$$+15 \quad 15$$
$$x < 40$$

2. Solution:

$$3x - 6 < 30$$
$$+6 \quad +6$$
$$3x < 36$$
$$x < 12$$

3. Solution:

$$3(x - 2) \geq 18$$
$$3x - 6 \geq 18$$
$$+6 \quad +6$$
$$3x \geq 24$$
$$x \geq 8$$

4. Solution:

$$4(2x - 5) \geq 3(2x + 10)$$
$$8x - 20 \geq 6x + 30$$
$$-6x \quad -6x$$
$$2x - 20 \geq 30$$
$$+20 + 20$$
$$2x \geq 50, \ x \geq 25$$

5. Solution:

$$-10 < 2x - 12 < 20$$
$$+12 \qquad +12 + 12$$
$$2 < 2x < 32$$
$$1 < x < 16$$

6. Solution:

$$8 \leq 4X - 20 \leq 16$$
$$+20 \qquad +20 + 20$$
$$28 \leq 4x \leq 36$$
$$7 \leq x \leq 9$$

7. Solution:

$$10x - 15(2x + 1) > 25$$
$$10x - 30x - 15 > 25$$
$$-20x - 15 > 25$$
$$+15\ +15$$
$$-20x > 40,\ x < -2$$

8. Solution:

$$12x - 3(x - 5) \le 42$$
$$12x - 3x + 15 \le 42$$
$$9x + 15 \le 42$$
$$-15\ -15$$
$$9x \le 27,\ x \le 3$$

9. Solution:

$$\frac{x + 1}{3} \le \frac{x - 3}{4} \quad (\text{cross} - \text{multiply})$$
$$4(x + 1) \le 3(x - 3)$$
$$4x + 4 \le 3x - 9$$
$$-3x \qquad -3x$$
$$x + 4 \le -9$$
$$-4\ -4$$
$$x \le -13$$

10. Solution:

$$\frac{x + 4}{x} \ge \frac{2}{3} \quad (\text{cross} - \text{multiply})$$
$$3(x + 4) \ge 2x$$
$$3x + 12) \ge 2x$$
$$-2x \qquad -2x$$
$$x + 12 \ge 0,\ x \ge -12$$

11. Solution:

$$7x - 8 \ge 3(x - 6)$$
$$7x - 8 \ge 3x - 18$$
$$-3x \qquad -3x$$
$$4x - 8 \ge -18$$
$$+8\ +8$$
$$4x \ge -10,\ x \ge \frac{-10}{4}$$
$$x \ge \frac{-5}{2}$$

12. Solution:

$$3x - (x + 5) > -2(2 + x) + 7$$
$$3x - x - 5 > -4 - 2x + 7$$
$$2x - 5 > 3 - 2x$$
$$2x \qquad +2x$$
$$4x - 5 > 3$$
$$+5\ +5$$
$$4x > 8,\ x > 2$$

13. Solution:

$$\frac{1}{3}(6x - 12 + 6) \leq 2x - 8$$

$$2x - 2 \leq 2x - 8$$

$$-2x \qquad -2x$$

$$+ \underline{\hspace{4cm}}$$

$$-2 \leq -8 \text{ (No solution)}$$

14. Solution:

$$3x - 5 + 4x \geq -4(x + 4)$$

$$7x - 5 \geq -4x - 16$$

$$4x \qquad + 4x$$

$$+ \underline{\hspace{4cm}}$$

$$11x - 5 \geq -16$$

$$+5 \quad +5$$

$$+ \underline{\hspace{4cm}}$$

$$11x \geq -11, \; x \geq -1$$

15. Solution:

$$x - 18 > -8$$

$$+18 \quad 18$$

$$+ \underline{\hspace{3cm}}$$

$$x > 10$$

16. Solution:

$$5x - 10 < 30$$

$$+10 \quad 10$$

$$+ \underline{\hspace{3cm}}$$

$$5x < 40, \; x < 8$$

17. Solution:

$$-3x - 18 + 20 \leq 2(x - 8)$$

$$-3x + 2 \leq 2x - 16$$

$$+3x \qquad 3x$$

$$+ \underline{\hspace{5cm}}$$

$$2 \leq 5x - 16$$

$$16 \qquad +16$$

$$+ \underline{\hspace{5cm}}$$

$$18 \leq 5x$$

$$\frac{18}{5} \leq x$$

18. Solution:

$$-7x - 14 \geq -14 - 2(x - 10)$$

$$+14 \qquad +14$$

$$+ \underline{\hspace{5cm}}$$

$$-7x \geq -2(x - 10)$$

$$-7x \geq -2x + 20$$

$$+2x \quad +2x$$

$$\underline{\hspace{5cm}}$$

$$-5x \geq 20, \; x \leq -4$$

SOLVING LITERAL EQUATIONS
EXERCISES SOLUTIONS

1. Solution:

$$a(b + c) = d$$

$$ab + ac = d$$

$$\underline{\begin{array}{cc} -ab & -ab \\ + & \end{array}}$$

$$\frac{ac}{a} = \frac{d - ab}{a}$$

$$c = \frac{d - ab}{a}$$

2. Solution:

$$\frac{1}{a} = \frac{1}{b} + \frac{1}{c}, \quad \frac{1 \cdot c}{a \cdot c} - \frac{1 \cdot a}{c \cdot a} = \frac{1}{b}$$

$$\frac{c}{ac} - \frac{a}{ac} = \frac{1}{b}, \quad \frac{c - a}{ac} = \frac{1}{b}, \text{ then } b = \frac{ac}{c - a}$$

3. Solution:

$$4x - 3y = 6$$

$$4x = 3y + 6$$

$$x = \frac{3y + 6}{4}$$

4. Solution:

$$V = \pi r^2 h, \quad \sqrt{\frac{V}{\pi h}} = \sqrt{r^2}$$

$$r = \sqrt{\frac{V}{\pi h}}$$

5. Solution:

$$P = 2L + 2W$$

$$\frac{P - 2L}{2} = \frac{2W}{2}, \quad W = \frac{P - 2L}{2}$$

$$W = \frac{P}{2} - L$$

6. Solution:

$$y = mx + b$$

$$\underline{\begin{array}{cc} -b & -b \\ + & \end{array}}$$

$$y - b = mx$$

$$\frac{y - b}{m} = x$$

7. Solution:

$$x = 2(y - 3x)$$

$$x = 2y - 6x$$

$$\underline{\begin{array}{cc} +6x & +6x \\ + & \end{array}}$$

$$\frac{7x}{2} = \frac{2y}{2}, \quad y = \frac{7x}{2}$$

8. Solution:

$$xy + 3 = m$$

$$\frac{xy}{x} = \frac{m - 3}{x}, \quad y = \frac{m - 3}{x}$$

9. Solution:

$$\frac{x+a}{3} = \frac{x-a}{4} \quad \text{(cross – multiply)}$$

$$4(x + a) = 3(x - a)$$

$$4x + 4a = 3x - 3a$$

$$\underline{ -3x \qquad -3x}$$

$$+$$

$$x + 4a = -3a, \quad x = -7a$$

$$\frac{-x}{7} = a$$

10. Solution:

$$R = \frac{R_1 \cdot R_2}{R_1 + R_2} \quad \text{(cross – multiply)}$$

$$R_1 \cdot R + R_2 \cdot R = R_1 \cdot R_2$$

$$-R_1 \cdot R \qquad\qquad -R_1 \cdot R$$

$$+$$

$$R_2 \cdot R = R_1 \cdot R_2 - R_1 \cdot R$$

$$\frac{R_2 \cdot R}{R_2 - R} = \frac{R_1 (R_2 - R)}{R_2 - R}$$

$$\frac{R_2 \cdot R}{R_2 - R} = R_1$$

11. Solution:

$$\frac{x}{3} = \frac{y}{4} \quad \text{(cross – multiply)}$$

$$3y = 4x$$

$$y = \frac{4}{3}x$$

12. Solution:

$$A = \frac{1}{2}bh \quad \text{(cross – multiply)}$$

$$\frac{2A}{b} = \frac{b \cdot h}{b}$$

$$\frac{2A}{b} = h$$

13. Solution:

$$E = mc^2, \quad c^2 = \frac{E}{m}$$

$$\sqrt{c^2} = \sqrt{\frac{E}{m}}, \quad c = \sqrt{\frac{E}{m}}$$

14. Solution:

$$ab = b + c$$

$$-b \quad -b$$

$$+$$

$$ab - b = c$$

$$\frac{b(a-1)}{a-1} = \frac{c}{a-1}, \quad b = \frac{c}{a-1}$$

15. Solution:

$$c = \frac{5}{9}(f - 32) \quad \text{(cross – multiply)}$$

$$9c = 5(f - 32)$$

$$9c = 5f - 160)$$

$$\underline{+160 \quad +160}$$

$$9c + 160 = 5f$$

$$\frac{9c + 160}{5} = f$$

$$\frac{9c}{5} + 32 = f$$

16. Solution:

$$x = \frac{1}{2}y - 2$$

$$\underline{+2 \qquad +2}$$

$$x + 2 = \frac{1}{2}y \quad \text{(cross – multiply)}$$

$$2(x + 2) = y$$

$$2x + 4 = y$$

ABSOLUTE VALUE EQUATIONS AND ABSOLUTE VALUE INEQUALITIES EXERCISES SOLUTIONS

1. Solution:

$|x - 5| = 12$, $x - 5 = \pm 12$

$$x - 5 = 12 \quad \text{or} \quad x - 5 = -12$$
$$+5 \ +5 \qquad\qquad +5 \ +5$$
$$x = 17 \quad \text{or} \quad x = -7$$

2. Solution:

$|2x - 8| = 24$, $2x - 8 = \pm 24$

$$2x - 8 = 24 \quad \text{or} \quad 2x - 8 = -24$$
$$+8 \ +8 \qquad\qquad +8 \ +8$$
$$2x = 32 \quad \text{or} \quad 2x = -16$$
$$x = 16 \quad \text{or} \quad x = -8$$

3. Solution:

$\left|\dfrac{x}{2} - 4\right| = 7$, $\dfrac{x}{2} - 4 = \pm 7$

$$\frac{x}{2} - 4 = 7 \quad \text{or} \quad \frac{x}{2} - 4 = -7$$
$$+4 \ +4 \qquad\qquad +4 \ +4$$
$$\frac{x}{2} = 11 \qquad\qquad \frac{x}{2} = -3$$
$$x = 22 \quad \text{or} \quad x = -6$$

4. Solution:

$\left|\dfrac{2}{x} + 3\right| = 5$, $\dfrac{2}{x} + 3 = \pm 5$

$$\frac{2}{x} + 3 = 5 \quad \text{or} \quad \frac{2}{x} + 3 = -5$$
$$-3 \ -3 \qquad\qquad -3 \ -3$$
$$\frac{2}{x} = 2 \qquad\qquad \frac{2}{x} = -8$$
$$2 = 2x \quad \text{or} \quad 2 = -8x$$
$$1 = x \quad \text{or} \quad \frac{-1}{4} = x$$

5. Solution:

$|x - 9| = |2x - 1|$

$$x - 9 = \pm(2x - 1) \qquad x - 9 = -(2x - 1)$$
$$x - 9 = 2x - 1 \quad \text{or} \quad x - 9 = -2x + 1$$
$$-x \qquad -x \qquad\qquad -x \qquad -x$$
$$-9 = x - 1 \quad \text{or} \quad -9 = -3x + 1$$
$$+1 \quad +1 \qquad\qquad -1 \qquad -1$$
$$-8 = x \quad \text{or} \quad \frac{10}{3} = x$$

6. Solution:

$\left|\dfrac{1}{2}x + 5\right| = 9$, $\dfrac{1}{2}x + 5 = \pm 9$

$$\frac{1}{2}x + 5 = 9 \quad \text{or} \quad \frac{1}{2}x + 5 = -9$$
$$-5 \ -5 \qquad\qquad -5 \ -5$$
$$\frac{1}{2}x = 4 \qquad\qquad \frac{1}{2}x = -14$$
$$x = 8 \quad \text{or} \quad x = -28$$

7. Solution:

$$\frac{3^3 \times 3^{10}}{(3^4)^3} = \frac{3^{13}}{3^{12}} = 3^{13-12} = 3^1$$

8. Solution:

$$\frac{2^4 \times 2^{12}}{(2^5)^3} = \frac{2^{16}}{2^{15}} = 2^{16-15} = 2^1$$

9. Solution:

$$\left|\frac{3x}{2} - 5\right| = 12, \quad \frac{3x}{2} - 5 = \pm 12$$

$$\frac{3x}{2} - \cancel{5} = 12 \quad \text{or} \quad \frac{3x}{2} - \cancel{5} = -12$$
$$+\cancel{5} \quad +5 \qquad\qquad +\cancel{5} \quad +5$$

$$\frac{3x}{2} = 17 \qquad\qquad \frac{3x}{2} = -7$$

$$3x = 34 \qquad\qquad 3x = -14$$

$$x = \frac{34}{3} \quad \text{or} \quad x = \frac{-14}{3}$$

10. Solution:

$$|x + 5| = -9 \text{ No Sulation}$$

11. Solution:

$$|x - 9| \le 12, \quad -12 \le x - \cancel{9} \le 12$$
$$+9 \quad +\cancel{9} \quad +9$$
$$-3 \le x \le 21$$

12. Solution:

$$\left|\frac{1}{3}x + 3\right| \le 9, \quad -9 \le \frac{1}{3}x + \cancel{3} \le 9$$
$$-3 \qquad -\cancel{3} \ -3$$
$$+$$

$$-12 \le \frac{1}{3}x \le 6$$

$$-36 \le x \le 18$$

13. Solution:

$|3x - 5| \le 5x, \; -5x \le 3x - 5 \le 5x$

$$-5x \le 3x - 5 \qquad \text{or} \qquad 3x - 5 \le 5x$$

$$\underline{-3x \quad -3x} \qquad\qquad \underline{-3x \qquad -3x}$$

$$-8x \le -5 \qquad\qquad\qquad -5 \le 2x$$

$$x \ge \frac{5}{8} \qquad \text{or} \qquad -\frac{5}{2} \le x$$

14. Solution:

$$\left|\frac{1}{2}x\right| \le 9, \; -9 \le \frac{1}{2}x \le 9$$

$$-18 \le x \le 18$$

15. Solution:

$$\left|\frac{3x}{2} + 5\right| - 8 = 18$$

$$\underline{\qquad\qquad +8 \quad +8}$$

$$\left|\frac{3x}{2} + 5\right| = 26, \; \frac{3x}{2} + 5 = \pm 26$$

$$\frac{3x}{2} + 5 = 26 \qquad \text{or} \qquad \frac{3x}{2} + 5 = -26$$

$$\underline{\quad -5 \quad -5} \qquad\qquad \underline{\quad -5 \quad -5}$$

$$\frac{3x}{2} = 21 \qquad\qquad\qquad \frac{3x}{2} = -31$$

$$3x = 42 \qquad\qquad\qquad 3x = -62$$

$$x = 14 \qquad \text{or} \qquad x = \frac{-62}{3}$$

16. Solution:

$$\left|\frac{2}{3}x - 5\right| < 12, \; -12 < \frac{2x}{3} - 5 < 12$$

$$-12 < \frac{2x}{3} - 5 < 12$$

$$\underline{+5 \qquad\quad +5 \quad +5}$$

$$-7 < \frac{2}{3}x < 17$$

$$-21 < 2x < 51$$

$$-\frac{21}{2} < x < \frac{51}{2}$$

1. Solution:

$$\frac{x+4}{3} = \frac{x-5}{4} \quad \text{(cross – multiply)}$$

$$4(x + 4) = 3(x - 5)$$

$$4x + 16 = 3x - 15$$

$$-3x \qquad -3x$$

$$+ \overline{\qquad\qquad\qquad}$$

$$x + 16 = -15$$

$$-16 \quad -16$$

$$+ \overline{\qquad\qquad\qquad}$$

$$x = -31$$

2. Solution:

$$\frac{x-8}{5} = \frac{x+4}{3} \quad \text{(cross – multiply)}$$

$$3(x - 8) = 5(x + 4)$$

$$3x - 24 = 5x + 20$$

$$-3x \qquad -3x$$

$$+ \overline{\qquad\qquad\qquad}$$

$$-24 = 2x + 20$$

$$-20 \qquad -20$$

$$+ \overline{\qquad\qquad\qquad}$$

$$-44 = 2x$$

$$-22 = x$$

3. Solution:

$$\frac{\frac{x}{2}}{\frac{1}{2}} = \frac{x-6}{2}$$

$$\frac{x}{2} \cdot \frac{2}{1} = \frac{x-6}{2}$$

$$\frac{x}{1} = \frac{x-6}{2} \quad \text{(cross – multiply)}$$

$$2x = x - 6$$

$$-x \quad -x$$

$$+ \overline{\qquad\qquad\qquad}$$

$$x = -6$$

4. Solution:

$$\frac{x}{5} = \frac{2x+4}{7} \quad \text{(cross – multiply)}$$

$$7x = 5(2x + 4)$$

$$7x = 10x + 20$$

$$-7x \quad -7x$$

$$+ \overline{\qquad\qquad\qquad}$$

$$0 = 3x + 20$$

$$-20 \qquad -20$$

$$+ \overline{\qquad\qquad\qquad}$$

$$-20 = 3x, \quad x = \frac{-20}{3}$$

5. Solution:

$$\frac{x+10}{2} = \frac{3x-5}{4} \text{ (cross – multiply)}$$

$$4(x + 10) = 2(3x - 5))$$

$$\cancel{4x} + 40 = 6x - 10$$

$$-\cancel{4x} \qquad - 4x$$

$$40 = 2x - \cancel{10}$$

$$+10 \qquad +\cancel{10}$$

$$50 = 2x$$

$$25 = x$$

6. Solution:

$$\frac{2x-1}{3} = \frac{3x+1}{5} \text{ (cross – multiply)}$$

$$5(2x - 1) = 3(3x + 1))$$

$$10x - 5 = \cancel{9x} + 3$$

$$-9x \qquad -\cancel{9x}$$

$$x - \cancel{5} = 3$$

$$+\cancel{5} \qquad +5$$

$$x = 8$$

7. Solution:

Direct Variation

$$y = kx$$

$$20 = k \cdot 5$$

$$4 = k$$

$$y = 4x, \text{ since } x = 3$$

$$y = 4 \cdot 3 = 12$$

8. Solution:

$$y = \frac{k}{x} \rightarrow \text{Inverse variation}$$

$$60 = \frac{k}{12}, \; k = 60 \cdot 12 = 720$$

$$y = \frac{720}{x} \text{ since } x = 18$$

$$y = \frac{720}{18} = 40$$

9. Solution:

$$y = \frac{7}{x} \rightarrow y = \frac{k}{x}$$

Inverse variation

10. Solution:

$$y = 5x \rightarrow y = kx$$

Direct Variation

11. Solution:

From table it's direct variation, because when x is increase, y is increase as well. (y = kx)

12. Solution:

From table it's inverse variation, because when x is increase, y is decrease. $\left(y = \frac{k}{x}\right)$

1. Solution:

$4\,(2x + y = 20)$

$3x - 4y = 19$

$8x + 4y = 80$

$+\ \underline{3x - 4y = 19}$

$11x = 99,\ x = 9$

$2 \cdot 9 + y = 20$

$18 + y = 20,\ y = 2$

2. Solution:

$x + 2y = 15$

$+\ \underline{3x - 2y = 17}$

$4x = 32,\qquad x = 8$

$x + 2y = 15,\qquad 8 + 2y = 15$

$2y = 7,\qquad y = \dfrac{7}{2}$

3. Solution:

$3x + 2y = 10$

$-2\,(x + y = 14)$

$3x + 2y = 10$

$+\ \underline{-2x - 2y = -28}$

$x = -18$

$x + y = 14,\ -18 + y = 14$

$y = 32$

4. Solution:

$2x + 6y = 18$

$6\,(2x - y = 4)$

$2x + 6y = 18$

$+\ \underline{12x - 6y = 24}$

$14x = 42,\qquad x = 3$

$2x - y = 4,\qquad 2 \cdot 3 - y = 4$

$6 - y = 4,\qquad y = 2$

5. Solution:

$x - \dfrac{1}{2}y = 16$

$+\ \underline{x + \dfrac{1}{2}y = 6}$

$2x = 22\qquad x = 11$

$11 + \dfrac{1}{2}y = 6,\quad \dfrac{1}{2}y = -5,\quad y = -10$

6. Solution:

$7x - y = 118$

$-\ (2x - y = 8)$

$7x - y = 118$

$+\ \underline{-2x + y = -8}$

$5x = 110,\qquad x = 22$

$2x - y = 8,\qquad 2 \cdot 22 - y = 8$

$y = 36$

7. Solution:

$2y + 2 - 4y = 8$

$-2y + 2 = 8, \qquad -2y = 6,$

$y = -3$

$x - 4y = 8, \qquad x - 4(-3) = 8,$

$x = -4$

8. Solution:

$x + 2(x + 6) = 18 \qquad\qquad y = x + 6$

$x + 2x + 12 = 18 \qquad\qquad y = 2 + 6$

$3x + 12 = 18 \qquad\qquad\quad y = 8$

$3x = 6$

$x = 2$

9. Solution:

$3y + y = 20, \qquad 4y = 20$

$y = 5$

$x = 3y, \qquad\qquad x = 3 \cdot 5 = 15$

10. Solution:

$2x - 3y = 0 \qquad\qquad y = 12 - 2x$

$2x - 3(12 - 2x) = 0 \qquad 2x - 36 + 6x = 0$

$8x = 36$

$x = \dfrac{36}{8} = \dfrac{9}{2}$

$2 \cdot \dfrac{9}{2} + y = 12, \qquad 9 + y = 12,$

$y = 3$

11. Solution:

$3y - \dfrac{1}{2}\,y = 25, \qquad \dfrac{6y - y}{2} = 25$

$\dfrac{5y}{2} = 25 \qquad\qquad y = 10$

$x = 3y$

$x = 3 \cdot 10 = 30$

12. Solution:

$x = 6y$

$6y + 3y = 36, \qquad 9y = 36$

$y = 4$

$x = 6y, \qquad\qquad x = 6 \cdot 4 = 24$

13. Solution:

$7k - 5 = -4k - 16$

$11k = -16 + 5, \quad 11k = -11, \quad k = -1$

14. Solution:

$5x + 15 = 4x - 12$

$5x - 4x = -12 - 15, \qquad x = -27$

15. Solution:

$3\ell + 2w = 48$

$w = 2\ell + 3$

$2\ell + 2(2\ell + 3) = 48$

$2\ell + 4\ell + 6 = 48$

$\ell = 7$

16. Solution:

$4C + 2A = 48 \qquad\qquad 4C + 2(3C) = 48$

$C = \dfrac{1}{3}\,A, \quad A = 3C \qquad 4C + 6C = 48$

$\qquad\qquad\qquad\qquad\qquad 10C = 48$

$\qquad\qquad\qquad\qquad\qquad C = \$4.8$

SOLVING EQUATIONS INVOLVING PARALLEL AND PERPENDICULAR LINES EXERCISES SOLUTIONS

1. Solution:

$y = mx + b$

$2x - 3y = 12$

$2x - 12 = 3y$

$\dfrac{2}{3}x - 4 = y$

$m_1 = \dfrac{2}{3}$

$m_2 = \dfrac{3}{12} = \dfrac{1}{4}$ Neither.

2. Solution:

$y = \dfrac{-2}{3}x + 6$

$m_1 = \dfrac{-2}{3}$

$3x - 18 = 2y$

$\dfrac{3}{2}x - 9 = y$

$m_2 = \dfrac{3}{2}$

since $m_1 \cdot m_2 = -1$

lines are perpendicular.

3. Solution:

$y = mx + b$

$3x - 12 = 4y$

$\dfrac{3x}{4} - 3 = y$

$m_1 = \dfrac{3}{4}$

$m_2 = \dfrac{-4}{3}$

Since $m_1 \cdot m_2 = -1$, Perpendicular.

4. Solution:

$y = m + +b$ $\qquad$ $m_1 \cdot m_2 = -1$

$y = -x + 6 \longrightarrow m_1 = -1$

$y = x + 0 \longrightarrow m_2 = 1$

So, lines are perpendicular.

6. Solution:

$y = mx + b$

$3x - 6 = 6y,\quad \dfrac{3}{6}x - \dfrac{6}{6} = y$

$m_1 = \dfrac{3}{6} = \dfrac{1}{2}$

$y = -2x$

$m_2 = -2$

Since the product of the slopes is equal to -1, then lines are perpendicular.

5. Solution:

$y = mx + b$

$m_1 = -2,\ m_2 = -2$

since slopes are equal, lines are parallel.

7. Solution:

$m_1 = 4$ and $m_1 \parallel m_2$, so $m_2 = 4$

$y = mx + b$

$y = 4x + b$, $3 = 4(2) + b$, $b = -5$

$y = 4x - 5$

8. Solution:

$m = \dfrac{1}{2}$ and $m_1 \parallel m_2$, so $m_2 = \dfrac{1}{2}$

$y = \dfrac{1}{2}x + b$, $\qquad 5 = \dfrac{1}{2}(1) + b$

$\qquad\qquad 10 = 1 + 2b$, $2b = 9$

$\qquad\qquad b = \dfrac{9}{2}$

$\qquad\qquad y = \dfrac{1}{2}x + \dfrac{9}{2}$

9. Solution:

$m_1 = 5$ and $m_1 \parallel m_2$, so $m_2 = 5$

$y = 5x + b$

$-4 = 5(-1) + b$

$-4 = -5 + b$

$b = 1$

$y = 5x + 1$

10. Solution:

$m_1 = \dfrac{2}{3}$ and $m_1 \parallel m_2$, so $m_2 = \dfrac{2}{3}$

$y = \dfrac{2}{3}x + b$, $3 = \dfrac{2}{3}(1) + b$ $\qquad b = \dfrac{7}{3}$

$y = \dfrac{2}{3}x + \dfrac{7}{3}$

11. Solution:

$m_1 = -1$

Since $m_1 \cdot m_2 = -1$, then $m_2 = 1$

$y = x + b$

$-3 = 2 + b$

$b = -5$

$y = x - 5$

12. Solution:

$m_1 = \dfrac{3}{2}$,

Since $m_1 \cdot m_2 = -1$

$m_2 = \dfrac{-2}{3}$

$y = \dfrac{-2}{3}x + b$, $\qquad -4 = \dfrac{-2}{3}(-1) + b$,

$b = \dfrac{-14}{3}$ $\qquad y = \dfrac{-2}{3}x - \dfrac{14}{3}$

13. Solution:

$y = x + a$

$m_1 = -1$

since $m_1 \cdot m_2 = -1$, $m_2 = 1$

$y = x + b$

$8 = 4 + b$

$4 = b$

$y = x + 4$

14. Solution:

$y = \dfrac{1}{2}x - \dfrac{5}{2}$ $\qquad m_1 = \dfrac{1}{2}$

Since $m_1 \cdot m_2 = -1$, then $m_2 = -2$

$y = -2x + b$, $\qquad -2 = -2(-1) + b$

$\qquad\qquad -2 = 2 + b$, $b = -4$

$y = -2x - 4$

1. Solution:

$x^2 - 6x + 3x^2 - 8x$

$= 4x^2 - 14x$

2. Solution:

$4x^2 - 3x + x^2 - 9x$

$= 5x^2 - 12x$

3. Solution:

$x^2 - x + 3 + x^2 - 8x + 10$

$= 2x^2 - 9x + 13$

4. Solution:

$x^2 - 3x + 15 + x^2 - 10x$

$= 2x^2 - 13x + 15$

5. Solution:

$x^2 - 9x + 5x^2 - 18x - 25$

$= 6x^2 - 27x - 25$

6. Solution:

$\dfrac{1}{2}x^2 + 2x + 5 + \dfrac{3}{2}x^2 - 8$

$= \dfrac{4}{2}x^2 + 2x + 5 - 8$

$= 2x^2 + 2x - 3$

7. Solution:

$x^3 - 9x^2 + 8x - 1 + x^2 - 18x$

$= x^3 - 8x^2 - 10x - 1$

8. Solution:

$x^4 + x^3 + 7 + x^2 - 12$

$= x^4 + x^3 + x^2 - 5$

9. Solution:

$(x^3 - x^2 + 8x - 1) - (x^2 - x)$

$= x^3 - x^2 + 8x - 1 - x^2 + x$

$= x^3 - 2x^2 + 9x - 1$

10. Solution:

$x^2 + 7 - (x^2 - x - 1)$

$= x^2 + 7 - x^2 + x + 1$

$= 8 + x$

11. Solution:

$(-x^2 + x - 1) - (2x^2 - 3x + 4)$

$= -x^2 + x - 1 - 2x^2 + 3x - 4$

$= -3x^2 + 4x - 5$

12. Solution:

$(x^4 + 7x + 5) - (x^2 - 10x + 3)$

$= x^4 + 7x + 5 - x^2 + 10x - 3$

$= x^4 - x^2 + 17x + 2$

13. Solution:

$\dfrac{1}{2} x^3 + 3 - (x^2 - 1)$

$\dfrac{1}{2} x^3 + 3 - x^2 + 1$

$\dfrac{1}{2} x^3 - x^2 + 4$

14. Solution:

$\left(\dfrac{3}{2}x^2 + 2x + 5\right) - \left(\dfrac{3}{2}x^2 + 5x + 12\right)$

$= \dfrac{3}{2}x^2 + 2x + 5 - \dfrac{3}{2}x^2 + 5x - 12$

$= 7x - 7$

15. Solution:

$(3x^2 + 3x^2) - (6x^2 - x)$

$= 6x^2 - 6x^2 + x$

$= x$

16. Solution:

$6x^2 - 2x - (2x - 5x^2)$

$= 6x^2 - 2x - 2x + 5x^2$

$= 11x^2 - 4x$

17. Solution:

$12x - x^3 - (x^3 - 3x)$

$= 12x - x^3 - x^3 + 3x$

$= -2x^3 + 15x$

18. Solution:

$x^4 - 4x - (x - 2x^4)$

$= x^4 - 4x - x + 2x^4$

$= 3x^4 - 5x$

1. Solution:

$(x - 1) \cdot (x + 1)$

$= x \cdot x + x \cdot 1 - 1 \cdot x - 1 \cdot 1$

$= x^2 + x - x - 1$

$= x^2 - 1$

2. Solution:

$(x - 4) \cdot (x + 5)$

$= x \cdot x + x \cdot 5 - 4 \cdot x - 4 \cdot 5$

$= x^2 + 5x - 4x - 20$

$= x^2 + x - 20$

3. Solution:

$(2x + 1)(3x - 5)$

$= 2x \cdot 3x + 2x \cdot (-5)\ 1 \cdot 3x + 1 \cdot (-5)$

$= 6x^2 - 10x + 3x - 5$

$= 6x^2 - 7x - 5$

4. Solution:

$(x^2 - 2)(x^2 + 2)$

$= x^2 \cdot x^2 + x^2 \cdot 2 - 2 \cdot x^2 - 2 \cdot 2$

$= x^4 + 2x^2 - 2x^2 - 4$

$= x^4 - 4$

5. Solution:

$\left(\dfrac{1}{2}x + 1\right)\left(\dfrac{1}{2}x - 1\right)$

$= \dfrac{1}{2}x \cdot \dfrac{1}{2}x + \dfrac{1}{2}x \cdot (-1) + 1 \cdot \dfrac{1}{2}x + 1 \cdot (-1)$

$= \dfrac{1}{4}x^2 - \dfrac{1}{2}x + \dfrac{1}{2}x - 1$

$= \dfrac{1}{4}x^2 - 1$

6. Solution:

$(x - 7) \cdot (x + 8)$

$= x \cdot x + x \cdot 8 - 7 \cdot x - 7 \cdot 8$

$= x^2 + 8x - 7x - 56$

$= x^2 + x - 56$

7. Solution:

$$(x) \cdot \left(\frac{1}{x} - 5\right)$$

$$= x \cdot \frac{1}{x} - x \cdot 5$$

$$= 1 - 5x$$

8. Solution:

$$x \, (x + 2)$$

$$= x^2 + 2x$$

9. Solution:

$$-4x \, (3x - 5)$$

$$= -4x \cdot 3x - 4x \cdot (-5)$$

$$= -12x^2 + 20x$$

10. Solution:

$$5x^2 \, (x^2 + 2x - 1)$$

$$= 5x^2 \cdot x^2 + 5x^2 \cdot 2x + 5x^2 \, (-1)$$

$$= 5x^4 + 10x^3 - 5x^2$$

11. Solution:

$$\frac{25x^2 y^3}{xy} = 25x^{\,2-1} y^{\,3-1}$$

$$= 25x^{\,1} y^2$$

12. Solution:

$$\frac{x^7 y^{12}}{2x^4 y^5} = \frac{1}{2} x^{\,7-4} y^{\,12-5}$$

$$= \frac{1}{2} x^{\,3} y^7$$

13. Solution:

$$\frac{25x^{-8}y^{10}}{5x^{10}y^{-5}} = \frac{5x^{-8}y^{10}}{x^{10}y^{-5}}$$

$$= 5x^{-8-10}y^{10-(-5)}$$

$$= 5x^{-18}y^{15}$$

$$= \frac{5y^{15}}{x^{18}}$$

14. Solution:

$$\frac{-33x^{20}y^{13}}{3x^{8}y^{7}}$$

$$= -11x^{20-8}y^{13-7}$$

$$= -11x^{12}y^{6}$$

15. Solution:

$$\frac{4x^{0}y^{0}}{8x^{1}y^{-1}} = \frac{1}{2} \cdot \frac{1}{xy^{-1}}$$

$$= \frac{1}{2x} \cdot \frac{y^{1}}{1} = \frac{y}{2x}$$

16. Solution:

$$\frac{-2x^{2}y^{3}}{x^{2}y^{3}} = -2x^{2-2}y^{3-3}$$

$$= -2 \cdot 1$$

$$= -2$$

17. Solution:

$$5x^{2}y(x^{2} - y^{2})$$

$$= 5x^{2}y \cdot (x^{2}) + 5x^{2}y \cdot (-y^{2})$$

$$= 5x^{4}y + 5x^{2}y^{3}$$

18. Solution:

$$(5x^{2}y^{4})(4x^{2}y^{8})$$

$$= 5 \cdot 4 \, x^{2} \cdot x^{2} \cdot y^{4} \, y^{8}$$

$$= 20x^{4}y^{12}$$

1. Solution:

$x^2 - 81x$

$= x(x - 81)$

2. Solution:

$16x^2 - 4x$

$= 4x(4x - 1)$

3. Solution:

$8x^3 - 64x$

$= 8x(x^2 - 8)$

4. Solution:

$25x^3 - 5x$

$= 5x(5x^2 - 1)$

$$a^2 - b^2 = (a - b) \cdot (a + b)$$

$= 5x(x\sqrt{5} - 1) \cdot (x\sqrt{5} + 1)$

5. Solution:

$27x^3 - 9x$

$= 9x(3x^2 - 1)$

$$a^2 - b^2 = (a - b) \cdot (a + b)$$

$= 9x(x\sqrt{3} - 1) \cdot (x\sqrt{3} + 1)$

6. Solution:

$x^3 + 2x^2 + x$

$= x(x^2 + 2x + 1)$

$= x(x + 1)^2$

7. Solution:

$x^2 - 5x + 6$

$x \quad x \quad -3 \quad -2$

$= (x - 3) \cdot (x - 2)$

8. Solution:

$x^2 + 5x + 6$

$x \quad x \quad 3 \quad 2$

$= (x + 3) \cdot (x + 2)$

9. Solution:

$x^2 - 11x + 30$

$x \quad x \quad -5 \quad -6$

$= (x - 5) \cdot (x - 6)$

10. Solution:

$x^2 + x - 72$

$x \quad x \quad -8 \quad 9$

$= (x - 8) \cdot (x + 9)$

11. Solution:

$$x^2 - x - 20$$

$$x \quad x \quad -5 \quad +4$$

$$= (x - 5) \cdot (x + 4)$$

12. Solution:

$$x^2 + 2x - 80$$

$$x \quad x \quad -8 \quad 10$$

$$= (x - 8) \cdot (x + 10)$$

13. Solution:

$$6x^3 + 8x^2 + 3x + 4$$
$$= (6x^3 + 8x^2) + (3x + 4)$$
$$= 2x^2(3x + 4) + 1(3x + 4)$$
$$= (3x + 4)(2x^2 + 1)$$
$$= (3x + 4) \cdot (2x^2 + 1)$$

14. Solution:

$$x^3 + 3x^2 - x - 3$$
$$= (x^3 + 3x^{2)} - (x + 3)$$
$$= x^2(3x + 4) - 1(x + 3)$$

$$a^2 - b^2 = (a - b) \cdot (a + b)$$

$$= (x + 3)[x^2 - 1]$$
$$= (x + 3)(x - 1) \cdot (x + 1)$$

15. Solution:

$$2x^3 + x^2 + 2x + 1$$
$$= (2x^3 + x^2) + (2x + 1)$$
$$= x^2(2x + 1) + 1(2x + 1)$$
$$= (x^2 + 1)(2x + 1)$$

16. Solution:

$$x^3 + 4x^2 - 3x - 12$$
$$= (x^3 + 4x^2) - (3x + 12)$$
$$= x^2(x + 4) - 3(x + 4)$$
$$= (x + 4) \cdot [x^2 - 3]$$
$$= (x + 4) \cdot (x - \sqrt{3}) \cdot (x + \sqrt{3})$$

17. Solution:

$$x^3 + x^2 - x - 1$$
$$= (x^3 + x^2) - (x + 1)$$
$$= x^2(x + 1) - 1(x + 1)$$

$$a^2 - b^2 = (a - b) \cdot (a + b)$$

$$= (x + 1)[x^2 - 1]$$
$$= (x + 1)(x - 1)(x + 1)$$

18. Solution:

$$x^3 - 4x^2 - 8x + 32$$
$$= (x^3 - 4x^2) - 8(x - 4)$$
$$= x^2(x - 4) - 8(x - 4)$$
$$= (x^2 - 8)(x - 4)$$
$$= (x - 2\sqrt{2}) \cdot (x + 2\sqrt{2})(x - 4)$$

1. Solution:

$$\frac{4 - 16x^2}{1 + 2x} = \frac{4(1 - 4x^2)}{1 + 2x}$$

$$= \frac{4(1 - 2x)(1 + 2x)}{(1 + 2x)}$$

$$= 4(1 - 2x)$$

2. Solution:

$$\frac{x - 2}{x^2 - 4} = \frac{x - 2}{(x - 2)(x + 2)} = \frac{1}{x + 2}$$

4. Solution:

$$\frac{x^2 - 9}{x^4 - 81} = \frac{(x - 3)(x + 3)}{x^4 - 81}$$

$$= \frac{(x - 3)(x + 3)}{(x^2 - 9)(x^2 + 9)}$$

$$= \frac{(x - 3)(x + 3)}{(x - 3)(x + 3)(x^2 + 9)} = \frac{1}{(x^2 + 9)}$$

3. Solution:

$$\frac{2x - 6}{x - 3} = \frac{2(x - 3)}{x - 3} = 2$$

5. Solution:

$$\frac{9x^2 - 25}{3x - 5} = \frac{(3x - 5)(3x + 5)}{3x - 5}$$

$$= 3x + 5$$

6. Solution:

$$\frac{7x - 28}{x - 4} = \frac{7(x - 4)}{x - 4} = 7$$

8. Solution:

$$\frac{x^2 - 6x}{x^2 - 36} = \frac{x(x - 6)}{(x - 6)(x + 6)}$$

$$= \frac{x}{x + 6}$$

7. Solution:

$$\frac{9x^2}{3x - 6} = \frac{\overset{3}{9}x^2}{\underset{1}{3}(x - 2)} = \frac{3x^2}{x - 2}$$

10. Solution:

$$\frac{x^2 - x - 6}{x^2 - 7x + 12} = \frac{(x - 3)(x + 2)}{(x - 3)(x - 4)}$$

$$= \frac{x + 2}{x - 4}$$

9. Solution:

$$\frac{(x - 2)(x + 4)}{(x + 2)(x + 4)} = \frac{x - 2}{x + 2}$$

11. Solution:

$$\frac{x^2 - 3}{x - \sqrt{3}} = \frac{(x - \sqrt{3})(x + \sqrt{3})}{(x - \sqrt{3})}$$

$$= x + \sqrt{3}$$

12. Solution:

$$\frac{x^4 - x^2}{x^2 - 1} = \frac{x^2(x^2 - 1)}{x^2 - 1}$$

$$= x^2$$

13. Solution:

$$\frac{x^3 + 2x^2 + x}{x + 1} = \frac{x(x^2 + 2x + 1)}{x + 1}$$

$$= \frac{x(x + 1)(x + 1)}{x + 1}$$

$$= x(x + 1)$$

14. Solution:

$$\frac{x - 5}{x^2 - 25} = \frac{x - 5}{(x - 5)(x + 5)}$$

$$= \frac{1}{(x + 5)}$$

15. Solution:

$$\frac{4x^2 - 1}{2x + 1} = \frac{(2x - 1)(2x + 1)}{2x + 1}$$

$$= 2x - 1$$

16. Solution:

$$\frac{2x^2 + 7x + 3}{x^2 + 2x - 3} = \frac{(2x + 1)(x + 3)}{(x + 3)(x - 1)}$$

$$= \frac{2x + 1}{x - 1}$$

17. Solution:

$$\frac{x^2 - 8}{x - 2\sqrt{2}} = \frac{(x - \sqrt{8})(x + \sqrt{8})}{x - 2\sqrt{2}}$$

$$= \frac{(x - 2\sqrt{2})(x + 2\sqrt{2})}{x - 2\sqrt{2}}$$

$$= x + 2\sqrt{2}$$

18. Solution:

$$\frac{x^2 - 25}{x^3 - 125} = \frac{(x - 5)(x + 5)}{x^3 - 5^3}$$

$$= \frac{(x - 5)(x + 5)}{(x - 5)(x^2 + 5x + 25)} = \frac{(x + 5)}{x^2 + 5x + 25}$$

1. Solution:

$$\frac{x}{x+2} + \frac{1}{x} = \frac{x(x)}{(x+2)\cdot(x)} + \frac{1(x+2)}{x(x+2)}$$

$$= \frac{x^2}{x^2+2x} + \frac{x+2}{x^2+2x} = \frac{x^2+x+2}{x^2+2x}$$

2. Solution:

$$\frac{x}{x-2} - \frac{1}{x} = \frac{x(x)}{(x-2)\cdot x} - \frac{1(x-2)}{x(x-2)}$$

$$= \frac{x^2}{x^2-2x} - \frac{x-2}{x^2-2x} = \frac{x^2-x+2}{x^2-2x}$$

3. Solution:

$$\frac{2x}{x-1} - \frac{3x}{x+1}$$

$$= \frac{2x(x+1)}{(x-1)\cdot(x+1)} - \frac{3x(x-1)}{(x+1)\cdot(x-1)}$$

$$= \frac{2x(x+1) - 3x(x-1)}{(x-1)\cdot(x+1)}$$

$$= \frac{2x^2+2x-3x^2+3x}{x^2-1} + \frac{-x^2+5x}{(x-1)\cdot(x+1)}$$

$$= \frac{x(5-x)}{(x-1)\cdot(x+1)}$$

4. Solution:

$$\frac{x+2}{x-2} - \frac{x-1}{x} = \frac{(x+2)x}{(x-2)\cdot x} - \frac{(x-1)1(x-2)}{x(x-2)}$$

$$= \frac{x^2+2x}{x^2-2x} - \frac{x^2-2x-x+2}{x^2-2x}$$

$$= \frac{x^2+2x-x^2+2x+x-2}{x^2-2x} = \frac{5x-2}{x^2-2x}$$

5. Solution:

$$\frac{x+2}{x-2} + \frac{x+3}{x-3} = \frac{(x+2)(x-3)}{(x-2)(x-3)} + \frac{(x+3)(x-2)}{(x-3)(x-2)}$$

$$= \frac{x^2-3x+2x-6}{x^2-3x-2x+6} + \frac{x^2-2x+3x-6}{x^2-2x-3x+6}$$

$$= \frac{x^2-x-6}{x^2-5x+6} + \frac{x^2+x-6}{x^2-5x+6}$$

$$= \frac{2x^2-12}{x^2-5x+6}$$

6. Solution:

$$\frac{\cancel{(x-2)}(x+2)}{\cancel{x-2}} - \frac{\cancel{(x-5)}(x+5)}{\cancel{x-5}}$$

$$x+2-(x+5)$$

$$x+2-x-5$$

$$=-3$$

7. Solution:

$$\frac{x^2-1}{x+1} + \frac{x-3}{x-3} = \frac{x^2-1}{x+1} + 1$$

$$= \frac{(x-1)(x+1)}{x+1} + 1 = x - 1 + 1$$

$$= x$$

8. Solution:

$$\frac{x^2+1}{x-1} - \frac{x^2-1}{x+1}$$

$$= \frac{(x^2+1)(x+1)}{(x-1)(x+1)} - \frac{(x^2-1)(x-1)}{(x+1)(x-1)}$$

$$= \frac{x^3+x^2+x+1}{x^2+x-x-1} - \frac{x^3-x^2-x+1}{x^2+x-x-1}$$

$$= \frac{x^3+x^2+x+1-x^3+x^2+x-1}{x^2-1}$$

$$= \frac{2x^2+2x}{x^2-1} = \frac{2x(x+1)}{(x-1)(x+1)} = \frac{2x}{x-1}$$

9. Solution:

$$\frac{x^2+2x+1}{x+1} + \frac{x^2+6x+9}{x+3}$$

$$= \frac{(x+1)(x+1)}{x+1} + \frac{(x+3)(x+3)}{x+3}$$

$$= x + 1 + x + 3$$

$$= 2x + 4$$

10. Solution:

$$\frac{x^2-x-12}{x-4} - \frac{x^2+2x-3}{x+3}$$

$$= \frac{(x-4)(x+3)}{x-4} - \frac{(x+3)(x-1)}{x+3}$$

$$= x + 3 - (x - 1)$$

$$= x + 3 - x + 1$$

$$= 4$$

11. Solution:

$$\frac{1}{x-4} + \frac{1}{x+3} = \frac{1(x+3)}{(x-4)(x+3)} + \frac{1(x-4)}{x+3(x-4)}$$

$$= \frac{x+3}{(x-4)(x+3)} + \frac{x-4}{(x+3)(x-4)}$$

$$= \frac{x+3+x-4}{(x-4)(x+3)} = \frac{2x-1}{(x-4)(x+3)}$$

12. Solution:

$$\frac{x+3}{x-4} - \frac{3x-1}{x(x-4)}$$

$$= \frac{(x+3)x}{(x-4)x} - \frac{3x-1}{x(x-4)}$$

$$= \frac{x^2+3x-3x+1}{x(x-4)} = \frac{x^2+1}{x(x-4)}$$

13. Solution:

$$\frac{4}{x-4} + \frac{5}{x-5}$$

$$= \frac{4(x-5)}{(x-4)(x-5)} + \frac{5(x-4)}{(x-5)(x-4)}$$

$$= \frac{4x - 20 + 5x - 20}{x^2 - 5x - 4x + 20}$$

$$= \frac{9x - 40}{x^2 - 9x + 20}$$

14. Solution:

$$\frac{x^2 - 81}{x - 9} - \frac{x^2 - 16}{x - 4}$$

$$= \frac{(x-9)(x+9)}{(x-9)} - \frac{(x-4)(x+4)}{(x-4)}$$

$$= x + 9 - (x + 4) = x + 9 - x - 4 = 5$$

16. Solution:

$$\frac{3x}{x-3} - \frac{x}{x+3}$$

$$= \frac{3x(x+3) - x(x-3)}{(x-3)(x+3)}$$

$$= \frac{3x^2 + 9x - x^2 + 3x}{x^2 + 3x - 3x - 9}$$

$$= \frac{2x^2 + 12x}{x^2 - 9}$$

15. Solution:

$$\frac{x-7}{x} + \frac{x-5}{2x}$$

$$= \frac{(x-7)(2)}{x(2)} + \frac{x-5}{2x}$$

$$= \frac{2x - 14 + x - 5}{2x} = \frac{3x - 19}{2x}$$

18. Solution:

$$\frac{x+1}{x^2-1} - \frac{x-1}{x^2+1}$$

$$= \frac{x+1}{(x-1)(x+1)} - \frac{x-1}{x^2+1}$$

$$= \frac{1(x^2+1)}{(x-1)(x^2+1)} - \frac{(x-1)(x-1)}{(x^2+1)(x-1)}$$

$$= \frac{x^2 + 1 - x^2 + x + x - 1}{x^3 - x^2 + x - 1}$$

$$= \frac{2x}{x^3 - x^2 + x - 1}$$

17. Solution:

$$\frac{x^3 + y^3}{x^2 - xy + y^2} + \frac{2x + 2y}{x + y}$$

$$= \frac{(x+y)(x^2 - xy + y^2)}{x^2 - xy + y^2} + \frac{2(x+y)}{x+y}$$

$$= x + y + 2$$

1. Solution:

$$\frac{x}{x+2} \cdot \frac{x^2-4}{x} = \frac{\cancel{x}}{\cancel{x+2}} \cdot \frac{(x-2)\cancel{(x+2)}}{\cancel{x}}$$

$$= x - 2$$

2. Solution:

$$\frac{x-3}{x-4} \cdot \frac{x^2-16}{x^2-9} = \frac{\cancel{x-3}}{\cancel{x-4}} \cdot \frac{\cancel{(x-4)}(x+4)}{\cancel{(x-3)}(x+3)}$$

$$= \frac{x+4}{x+3}$$

3. Solution:

$$\frac{7x}{\cancel{12}^{\,1}} \cdot \frac{\overset{3}{\cancel{36x}}}{\underset{2}{\cancel{14x^2}}}$$

$$= \frac{x}{1} \cdot \frac{3x}{2x^2} = \frac{3x^2}{2x^2} = \frac{3}{2}$$

4. Solution:

$$\frac{3x+30}{21x} \cdot \frac{7x+14}{x+10} = \frac{\overset{1}{\cancel{3}}(\cancel{x+10})}{\underset{\underset{1}{3}}{21x}} \cdot \frac{\overset{1}{\cancel{7}}(x+2)}{\cancel{x+10}}$$

$$= \frac{1}{\underset{x}{\cancel{3x}}} \cdot \frac{x+2}{1} = \frac{x+2}{x}$$

5. Solution:

$$\frac{1}{x-8} \cdot \frac{x^2-64}{x+8}$$

$$\frac{1}{\cancel{x-8}} \cdot \frac{\cancel{(x-8)}\cancel{(x+8)}}{\cancel{x+8}} = 1$$

6. Solution:

$$\frac{2x+3}{x-7} \cdot \frac{2x-14}{4x+6} = \frac{\cancel{2x+3}}{\cancel{x-7}} \cdot \frac{\cancel{2}\cancel{(x-7)}}{\cancel{2}(2\cancel{x+3})}$$

$$= 1$$

7. Solution:

$$\frac{x^3-y^3}{x^2+xy+y^2} \cdot \frac{x}{x-y}$$

$$= \frac{\cancel{(x-y)}\cancel{(x^2+xy+y^2)}}{\cancel{x^2+xy+y^2}} \cdot \frac{x}{\cancel{x-y}}$$

$$= x$$

8. Solution:

$$\frac{x^2-100}{x-10} \cdot \frac{x^2-25}{x-5}$$

$$\frac{\cancel{(x-10)}(x+10)}{\cancel{(x-10)}} \cdot \frac{\cancel{(x-5)}(x+5)}{\cancel{(x-5)}}$$

$$= (x+10) \cdot (x+5)$$

$$= x^2 + 5x + 10x + 50$$

$$= x^2 + 15x + 50$$

9. Solution:

$$\frac{4x}{x-6} \div \frac{4x}{2x-12}$$

$$\frac{4x}{x-6} \cdot \frac{2(x-6)}{4x}$$

$$= 2$$

10. Solution:

$$\frac{x^2-64}{x-8} \div \frac{x^2-49}{x+7}$$

$$= \frac{(x-8)(x+8)}{(x-8)} \cdot \frac{(x+7)}{(x-7)(x+7)}$$

$$= \frac{x+8}{x-7}$$

11. Solution:

$$\frac{x^2+2x+1}{x+1} \div \frac{x^2+6x+9}{x+3}$$

$$= \frac{(x+1)(x+1)}{(x+1)} \cdot \frac{x+3}{(x+3)(x+3)}$$

$$= \frac{x+1}{x+3}$$

12. Solution:

$$\frac{x^2-8x+15}{x^2-9} \div \frac{x^2-4x-5}{x^2+3x}$$

$$= \frac{(x-3)(x-5)}{(x-3)(x+3)} \cdot \frac{x(x+3)}{(x-5)(x+1)}$$

$$= \frac{x}{x+1}$$

13. Solution:

$$\frac{x^2-y^2}{x^2+xy} \div \frac{x^2-xy}{xy+x}$$

$$= \frac{(x-y)(x+y)}{x(x+y)} \cdot \frac{x(y+1)}{x(x-y)} = \frac{y+1}{x}$$

14. Solution:

$$\frac{x^2-y^2}{x-y} \div \frac{x^2+y^2}{x+y}$$

$$= \frac{(x-y)(x+y)}{x-y} \cdot \frac{(x+y)}{x^2+y^2}$$

$$= \frac{(x+y)(x+y)}{x^2+y^2} = \frac{x^2+2xy+y^2}{x^2+y^2}$$

15. Solution:

$$\frac{x^2 + 4x + 16}{x + 4} \div \frac{x^2 - 6x + 9}{x - 3}$$

$$= \frac{(x+4)(x+4)}{(x+4)} \cdot \frac{x-3}{(x-3)(x-3)} = \frac{x+4}{x-3}$$

16. Solution:

$$\frac{y}{y-1} \div \frac{y^2}{y^3 - y}$$

$$\frac{y}{y-1} \cdot \frac{y(y^2 - 1)}{y^2}$$

$$\frac{y}{y-1} \cdot \frac{y(y-1)(y+1)}{y^2}$$

$$= y + 1$$

17. Solution:

$$\frac{3x^2 + 5x - 2}{3x - 1} \div \frac{x + 2}{5}$$

$$= \frac{(3x-1)(x+2)}{(3x-1)} \cdot \frac{5}{x+2} = 5$$

18. Solution:

$$\frac{12x^2 - 14x - 6}{3x + 1} \div \frac{4x - 6}{x^2}$$

$$= \frac{12x^2 - 14x - 6}{3x + 1} \cdot \frac{x^2}{4x - 6}$$

$$= \frac{(3x+1)(4x-6)}{(3x+1)} \cdot \frac{x^2}{(4x-6)} = x^2$$

COMPLEX FRACTIONS EXERCISES SOLUTIONS

1. Solution:

$$\frac{\dfrac{1}{3}}{\dfrac{3}{7}} = \frac{1}{3} \cdot \frac{7}{3} = \frac{7}{9}$$

2. Solution:

$$\frac{\dfrac{x}{y}}{\dfrac{a}{b}} = \frac{x}{y} \cdot \frac{b}{a} = \frac{bx}{ay}$$

3. Solution:

$$\frac{\dfrac{x-1}{6}}{\dfrac{x-2}{18}} = \frac{x-1}{\cancel{6}} \cdot \frac{\cancel{18}^{\,3}}{x-2}$$

$$= \frac{3x-3}{x-2}$$

4. Solution:

$$\frac{\dfrac{x+1}{24}}{\dfrac{x-4}{48}} = \frac{x+1}{\cancel{24}} \cdot \frac{\cancel{48}^{\,2}}{x-4} = \frac{2x+2}{x-4}$$

5. Solution:

$$\frac{\dfrac{x-3}{5}}{\dfrac{x \cdot 3}{2 \cdot 3} - \dfrac{x \cdot 2}{3 \cdot 2}} \pm \frac{\dfrac{x-3}{5}}{\dfrac{3x-2x}{6}} = \frac{\dfrac{x-3}{5}}{\dfrac{x}{6}}$$

$$= \frac{x-3}{5} \cdot \frac{6}{x} = \frac{6x-18}{5x}$$

6. Solution:

$$\frac{\dfrac{x-2}{6} - \dfrac{(x+2) \cdot 2}{3 \cdot 2}}{\dfrac{x}{4}} = \frac{\dfrac{x-2}{6} - \dfrac{2x+4}{6}}{\dfrac{x}{4}}$$

$$= \frac{-x-6}{\cancel{6}_{3}} \cdot \frac{\cancel{4}^{\,2}}{x} = \frac{-2x-12}{3x}$$

7. Solution:

$$\frac{x}{\dfrac{x \cdot 5}{3 \cdot 5} - \dfrac{x \cdot 3}{5 \cdot 3}} = \frac{x}{\dfrac{5x-3x}{15}}$$

$$= \frac{x}{1} \cdot \frac{15}{2x} = \frac{15\cancel{x}}{2\cancel{x}} = \frac{15}{2}$$

8. Solution:

$$\frac{1+\dfrac{1}{x}}{1-\dfrac{2}{x}} = \frac{\dfrac{x+1}{x}}{\dfrac{x-2}{x}} = \frac{x+1}{\cancel{x}} \cdot \frac{\cancel{x}}{x-2}$$

$$= \frac{x+1}{x-2}$$

9. Solution:

$$\frac{\dfrac{a}{b}}{\dfrac{x}{1}} = \frac{a}{b} \cdot \frac{1}{x} = \frac{a}{bx}$$

10. Solution:

$$\frac{x}{\dfrac{3}{x} - \dfrac{5}{x}} = \frac{x}{\dfrac{-2}{x}} = \frac{x}{1} \cdot \frac{x}{-2}$$

$$= \frac{-x^2}{2}$$

11. Solution:

$$\frac{\dfrac{2}{1\dfrac{1}{x}}}{3\dfrac{3}{x}} = \frac{\dfrac{2}{\dfrac{x+1}{x}}}{\dfrac{3x+3}{x}}$$

$$= \frac{\dfrac{2x}{x+1}}{\dfrac{3x+3}{x}} = \frac{2x}{x+1} \cdot \frac{x}{3x+3}$$

$$= \frac{2x^2}{3x^2 + 3x + 3x + 3} = \frac{2x^2}{3x^2 + 6x + 3}$$

12. Solution:

$$\frac{\dfrac{x}{5}}{2 \cdot \dfrac{x}{4} - \dfrac{x}{8}} = \frac{\dfrac{x}{5}}{\dfrac{2x - x}{8}} = \frac{\dfrac{x}{5}}{\dfrac{x}{8}}$$

$$= \frac{\cancel{x}}{5} \cdot \frac{8}{\cancel{x}} = \frac{8}{5}$$

13. Solution:

$$\frac{\dfrac{2}{a}}{\dfrac{a}{6}} = \frac{2}{a} \cdot \frac{6}{a}$$

$$= \frac{12}{a^2}$$

14. Solution:

$$\frac{\dfrac{x}{5}}{\dfrac{10}{x}} = \frac{x}{5} \cdot \frac{x}{10} = \frac{x^2}{50}$$

FUNCTION NOTATION, DOMAIN, RANGE AND INVERSE FUNCTIONS EXERCISES SOLUTIONS

1. Solution:

Domain : 1, 5, 6

Yes it is a function because domain is not repait

2. Solution:

Domain : 1, 2, 3, 1

No it is not a function because domain is repait

3. Solution:

$f(x) = 2x - 8,$

$f(-2) = 2(-2) - 8 = -4 - 8 = -12$

4. Solution:

$f(x) = x^2 - 5$

$f(3) = 3^2 - 5 = 9 - 5 = 4$

5. Solution:

$f(x) = 2x^2 + 3$

$f\left(\dfrac{1}{2}\right) = 2\left(\dfrac{1}{2}\right)^2 + 3$

$= 2 \cdot \dfrac{1}{4} + 3 = \dfrac{1}{2} + 3$

$= \dfrac{7}{2}$

6. Solution:

$f(x) = x^2 + 6$

$f(7) = (7)^2 + 6 = 49 + 6 = 55$

7. Solution:

$f(x) = 2x^2 + 2x - 5$

$f(-3) = 2(-3)^2 + 2x - 5$

$f(-3) = 2(9) - 6 - 5$

$f(-3) = 7$

8. Solution:

$f(x) = x^2 - 10x + 12$

$f\left(\dfrac{-1}{2}\right) = \left(\dfrac{-1}{2}\right)^2 - 10 \cdot \left(\dfrac{-1}{2}\right) + 12$

$= \dfrac{1}{4} + 5 + 12 = \dfrac{1}{4} + 17 = \dfrac{69}{4}$

9. Solution:

$$h(x) = \frac{4x - 1}{3}$$

$$h(-1) = \frac{4(-1) - 1}{3}$$

$$= \frac{-4 - 1}{3} = \frac{-5}{3}$$

10. Solution:

$$h(x) = \frac{x^2 - 6x + 12}{3}$$

$$h(3) = \frac{3^2 - 6(3) + 12}{3}$$

$$= \frac{9 - 18 + 12}{3} = \frac{-9 + 12}{3}$$

$$= \frac{3}{3} = 1$$

11. Solution:

$$g(x) = 2x^2 + 4x + 5$$

$$h\left(\frac{x}{2}\right) = 2\left(\frac{x}{2}\right)^2 + 4\left(\frac{x}{2}\right) + 5$$

$$= \overset{1}{2} \cdot \frac{x^2}{\underset{2}{\cancel{4}}} + 2x + 5$$

$$= \frac{x^2}{2} + 2x + 5$$

12. Solution:

$$g(x) = x^2 - 6x$$

$$g(3x) = (3x)^2 - 6(3x)$$

$$= 9x^2 - 18x$$

13. Solution:

$$h(x) = x^2 + 12$$

$$h(\sqrt{x}) = (\sqrt{x})^2 + 12$$

$$= x + 12$$

14. Solution:

$$g(x) = 3x - 5$$

$$g(-3x) = 3(-3x) - 5$$

$$= -9x - 5$$

15. Solution:

$f(x) = x + 4$

$y = x + 4$

$y - 4 = x$ (change x to y for inverse)

$x - 4 = y^{-1}$

$x - 4 = f^{-1}(x)$

16. Solution:

$f(x) = 2x - 54$

$y = 2x - 54, \quad y + 54 = 2x$

$\dfrac{y - 54}{2} = \dfrac{2x}{2}$

$\dfrac{y + 54}{2} = x$ (change x to y for inverse)

$f^{-1}(x) = \dfrac{x + 54}{2}$

17. Solution:

$f(x) = x^2 - 3$

$y = x^2 - 3$

$\sqrt{y + 3} = \sqrt{x^2}$

$\sqrt{y + 3} = x$ (change x to y for inverse)

$f^{-1}(x) = \sqrt{x + 3}$

18. Solution:

$f(x) = \sqrt{x - 8}$

$y = \sqrt{x - 8}, \qquad y^2 = x - 8$

$y^2 + 8 = x$ (change x to y for inverse)

$f^{-1}(x) = x^2 + 8$

OPERATIONS WITH FUNCTIONS AND COMPOSITION OF FUNCTIONS EXERCISES SOLUTIONS

1. Solution:

$= -3x + 9 + 2x + 5$

$= -x + 14$

2. Solution:

$= -3x + 9 - (2x + 5)$

$= -3x + 9 - 2x - 5 = -5x + 4$

3. Solution:

$= x^2 + 3x + 9 + 2(x + 12)$

$= x^2 + 3x + 9 + 2x + 24$

$= x^2 + 5x + 33$

4. Solution:

$= 3(x^2 + 3x + 9) - (x + 12)$

$= 3x^2 + 9x + 27 - x - 12$

$= 3x^2 + 8x + 15$

5. Solution:

$= (3x + 10) + 4$

$= 3x + 14$

6. Solution:

$= 3(x + 4) + 10$

$= 3x + 12 + 10 = 3x + 22$

7. Solution:

$= 5(6x - 15)$

$= 30x - 75$

8. Solution:

$= 6(5x) - 15$

$= 30x - 15$

9. Solution:

$= g(3) = 3 - 3 = 0$

$= f(0) = 2 \cdot 0^2 + 1 = 1$

10. Solution:

$= f(5) = 2 \cdot 5^2 + 1 = 50 + 1 = 51$

$= g(51) = 51 - 3 = 48$

11. Solution:

$= g(-1) = 4(-1)^2 + 5(-1) - 12 = 4 - 5 - 12 = -13$

$= f(-13) = 2(-13)^2 + (-13) + 3$

$= 2 \cdot 169 - 13 + 3 = 328$

12. Solution:

$f(-3) = 2(-3)^2 + (-3) + 3$

$= 2 \cdot 9 + 0$

$= 18$

$g(18) = 4(18)^2 + 5 \cdot (18) - 12$

$= 1296 + 90 - 12 = 1374$

13. Solution:

$= (x^2 - x)(x^2 + x - 2)$

$= (x^4 + x^3 - 2x^2 - x^3 - x^2 + 2x)$

$= x^4 - 3x^2 + 2x$

14. Solution:

$= \dfrac{x^2 - x}{x^2 + x - 2} = \dfrac{x(x - 1)}{(x + 2)(x - 1)}$

$= \dfrac{x}{x + 2}$

1. $f(x) = 2^x$

$f(-2) = 2^{-2} = \dfrac{1}{4}$

2. $f(x) = 3^{x-2}$

$f(4) = 3^{4-2} = 3^2 = 9$

3. $f(x) = \dfrac{2x-5}{4}$

$f(-1) = \dfrac{2^{-1-5}}{4} = \dfrac{2^{-6}}{2^2} = 2^{-6-2} = 2^{-8} = \dfrac{1}{2^8}$

4.

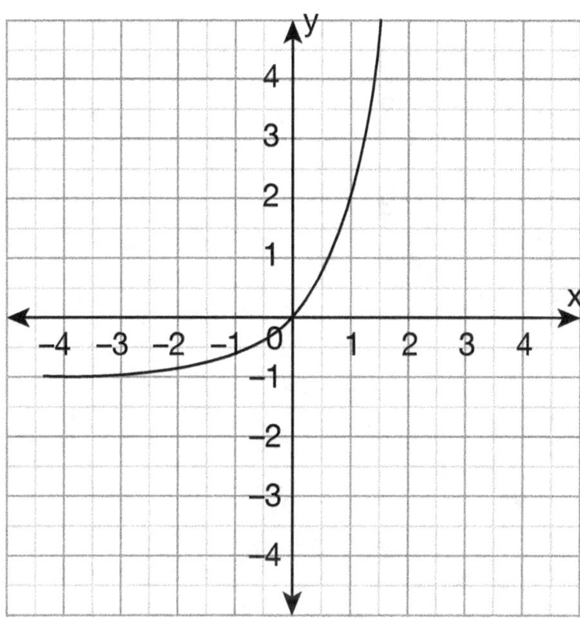

5.

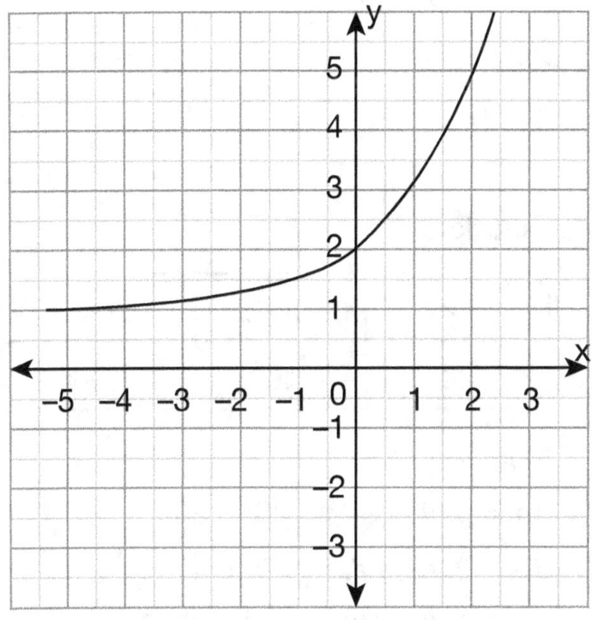

6. $3^{x-1} = 243$

$3^{x-1} = 3^5$, then $x - 1 = 5$ and $x = 6$

7. Solution:

$2^1 \cdot 2^{x+1} = 32$

$2^{x+2} = 2^5$

$x + 2 = 5$

$x = 3$

8. Solution:

$9^{2x-3} = 81^{3x-8}$

$(3^2)^{2x-3} = (3^4)^{3x-8}$

$2(2x-3) = 4(3x-8)$

$$2x - 3 = 6x - 16$$

$$\underline{+\quad -2x \qquad -2x}$$

$$-3 = 4x - 16$$

$$\underline{+\quad 16 \qquad +16}$$

$13 = 4x$, then $x = \dfrac{13}{4}$

9. Solution:

$\left(\dfrac{3}{2}\right)^{x-2} = \left(\dfrac{9}{4}\right)^{3x+2}$

$\left(\dfrac{3}{2}\right)^{x-2} = \left(\dfrac{3}{2}\right)^{2(3x+2)}$

$x - 2 = 6x + 4$

$-6 = 5x$

$\dfrac{-6}{5} = x$

10. Solution:

$8^{2x-3} = 64^{3x-8}$

$(2^3)^{2x-3} = (2^6)^{3x-2}$

$3(2x-3) = 6(3x-2)$

$2x - 3 = 6x - 4$

$1 = 4x \qquad\qquad \dfrac{1}{4} = x$

11. Solution:

From table each time x increase by 1, and y increase by a factor of 5. This is an exponential function. $y = 5^x$

$f(x) = 5^x$

When $x = 0$, $f(0) = 5^0 = 1$

When $x = 1$, $f(1) = 5^1 = 5$

When $x = 2$, $f(2) = 5^2 = 25$

When $x = 3$, $f(3) = 5^3 = 125$

12. Solution:

Day 1	Day 2	Day 3
20	400	8000

$f(x) = 20^x$

If $x = 1$, then $f(1) = 20^1 = 20$

If $x = 2$, then $f(2) = 20^2 = 400$

If $x = 3$, then $f(3) = 20^3 = 8000$

TRIGONOMETRIC FUNCTIONS EXERCISES SOLUTIONS

1) Solution:

$$\sin 120° = -\cos 30° = \frac{-\sqrt{3}}{2}$$

2) Solution:

$$\cos 135° = -\sin 45°$$

$$= \frac{-1}{\sqrt{2}} = \frac{-\sqrt{2}}{2}$$

3) Solution:

$$25° + 360° = 385° \text{ or } 25° - 360° = -335°$$

4) Solution:

$$\frac{\pi}{2} + 2\pi = \frac{5}{2}\pi \text{ or } \frac{\pi}{2} - 2\pi = \frac{-3\pi}{2}$$

5) Solution: From right triangle:

$$\sin 30° = \frac{op}{hyp} = \frac{1}{2}$$

6) Solution:

From $\dfrac{D}{180°} = \dfrac{R}{\pi}$, then $\dfrac{200}{180°} = \dfrac{R}{\pi}$

(cross–multiply)

$$R = \frac{200\pi}{180°} = \frac{20\pi}{18} = \frac{10\pi}{9}$$

7) Solution:

From $\dfrac{D}{180°} = \dfrac{R}{\pi}$, then $\dfrac{D}{180°} = \dfrac{\frac{\pi}{5}}{\pi}$

$$= \frac{D}{180°} = \frac{\pi}{5} \cdot \frac{1}{\pi}, \frac{D}{180°} = \frac{1}{5} \text{ (cross–multiply)}$$

$$D = \frac{180°}{5} = 36°$$

8) Solution:

Length of the Arc $= \dfrac{\pi r^2 \theta}{360°}$

$$= \frac{\pi \cdot 10^2 \cdot 90°}{360°} = \frac{100° \pi \cdot 90°}{360°} = \frac{100\pi}{4} = 25\pi$$

9) Solution:

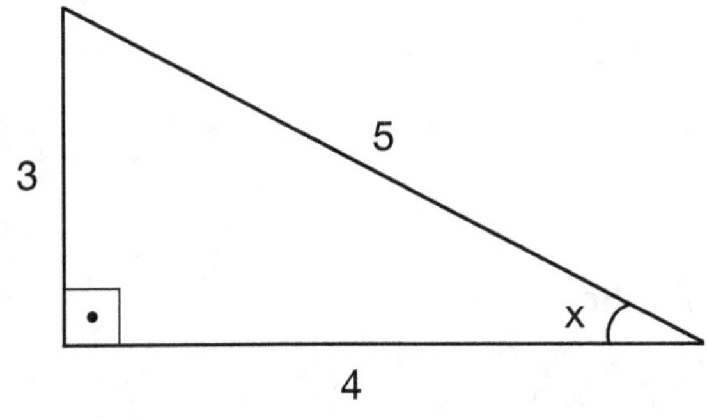

$$\cot x = \frac{Adj}{Opposite} \qquad \cot x = \frac{4}{3}$$

10) Solution:

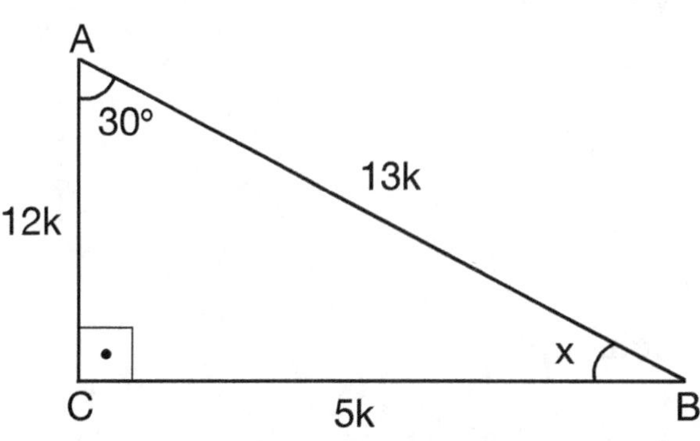

$$\tan(90 - x°) = \cot x° = \frac{5k}{12k} = \frac{5}{12}$$

1. Solution:

$$\sqrt{169} = \sqrt{13^2} = 13$$

2. Solution:

$$\sqrt{400} = \sqrt{20^2} = 20$$

3. Solution:

$$\sqrt{144} = \sqrt{12^2} = 12$$

4. Solution:

$$\sqrt{10000} = \sqrt[2]{10^4} = 100$$

5. Solution:

$$\sqrt{20} = \sqrt{4 \cdot 5} = \sqrt{2^2 \times 5} = 2\sqrt{5}$$

6. Solution:

$$\sqrt{1000} = \sqrt{100 \times 10}$$

$$\sqrt{10^2 \times 10} = 10\sqrt{10}$$

7. Solution:

$$\sqrt{x^4} = \sqrt[2]{x^4} = x^2$$

8. Solution:

$$\sqrt{y^{16}} = \sqrt[2]{y^{16}} = y^8$$

9. Solution:

$$\sqrt{25x^4} = \sqrt{5^2 x^4} = 5x^2$$

10. Solution:

$$\sqrt{36x^6} = \sqrt[2]{6^2 x^6} = 6x^3$$

11. Solution:

$$\sqrt{144x^4y^{10}} = \sqrt[2]{12^2 x^{4^2} y^{10^5}} = 12x^2y^5$$

12. Solution:

$$\sqrt{64x^8y^6} = \sqrt[2]{8^2 x^{8^4} y^{6^3}} = 8^1 x^4 y^3$$

13. Solution:

$$\sqrt{\frac{1}{25}x^4y^8} = \sqrt[2]{\frac{1}{5^2}x^4y^8} = \frac{1}{5}x^2y^4$$

14. Solution:

$$\sqrt{400x^2y^2z^2} = \sqrt{20^2 x^2 y^2 z^2} = 20xyz$$

15. Solution:

$$\sqrt{20x^2y^3} = \sqrt[2]{2^{2^1} x5x^{2^1} y^{2^1} y} = 2xy\sqrt{5y}$$

16. Solution:

$$\sqrt{\frac{18x^8}{3x^3}} = \sqrt{\frac{18x^8}{3x^3}} = \sqrt{\frac{18}{3}x^{8-3}}$$

$$= \sqrt{6\,x^5} = \sqrt[2]{6\,x^{4^2}x^1} = x^2\sqrt{6\,x}$$

17. Solution:

$$\sqrt{81x^{12}y^{18}z^{16}} = \sqrt[2]{9^{2^1} x^{12^6} y^{18^9} z^{16^8}}$$

$$= 9^1 x^6 y^9 z^8$$

18. Solution:

$$\sqrt{\frac{64x^6y^8}{16x^1y^1}} = \sqrt{\frac{64x^{6-1}y^{8-1}}{16}}$$

$$\sqrt[2]{4x^5y^7} = \sqrt{2^2 x^4 \cdot x^1 y^6}$$

$$= 2x^2 y^3 \sqrt[2]{xy}$$

ADDING AND SUBTRACTING RADICAL EXPRESSIONS EXERCISES SOLUTIONS

1. Solution:

$2\sqrt{2} + 3\sqrt{2}$

$= \sqrt{2}\,(2 + 3)$

$= 5\sqrt{2}$

2. Solution:

$6\sqrt{3} + 9\sqrt{3}$

$= \sqrt{3}\,(6 + 9)$

$= 15\sqrt{3}$

3. Solution:

$7\sqrt{2} - 3\sqrt{2}$

$= (7 - 3)\sqrt{2}$

$= 4\sqrt{2}$

4. Solution:

$12\sqrt{5} - 7\sqrt{5}$

$= (12 - 7)\sqrt{5}$

$= 5\sqrt{5}$

5. Solution:

$\sqrt{20} + \sqrt{18}$

$= \sqrt{4 \times 5} + \sqrt{9 \times 2}$

$= 2\sqrt{5} + 3\sqrt{2}$

6. Solution:

$\sqrt{24} + \sqrt{72}$

$= \sqrt{4 \times 6} + \sqrt{36 \times 2}$

$= 2\sqrt{6} + 6\sqrt{2}$

7. Solution:

$\sqrt{63} + \sqrt{28}$

$= \sqrt{7 \times 9} - \sqrt{4 \times 7}$

$= 3\sqrt{7} - 2\sqrt{7}$

$(3 - 2)\sqrt{7}$

$= \sqrt{7}$

8. Solution:

$\sqrt{48} - \sqrt{12}$

$= \sqrt{16 \times 3} - \sqrt{3 \times 4}$

$= 4\sqrt{3} - 2\sqrt{3}$

$= (4 - 2)\sqrt{3}$

$= 2\sqrt{3}$

9. Solution:

$6\sqrt{2} + \sqrt{32}$

$= 6\sqrt{2} + \sqrt{16 \times 2}$

$= 6\sqrt{2} + 4\sqrt{2}$

$= 10\sqrt{2}$

10. Solution:

$5\sqrt{4} + 3\sqrt{36}$

$= 5 \cdot 2 + 3 \cdot 6$

$= 10 + 18$

$= 28$

11. Solution:

$18\sqrt{2} - \sqrt{98}$

$= 18\sqrt{2} - \sqrt{49 \times 2}$

$= 18\sqrt{2} - 7\sqrt{2}$

$= 11\sqrt{2}$

12. Solution:

$48\sqrt{2} - 3\sqrt{128}$

$= 48\sqrt{2} - 3\sqrt{64 \times 2}$

$= 48\sqrt{2} - 3 \times 8\sqrt{2}$

$= 48\sqrt{2} - 24\sqrt{2} = 24\sqrt{2}$

14. Solution:

$4y\sqrt{2} + 3\sqrt{72y^2}$

$= 4y\sqrt{2} + 3\sqrt{36 \cdot 2y^2}$

$= 4y\sqrt{2} + 3 \cdot 6 \cdot y\sqrt{2}$

$= 4y\sqrt{2} + 18y\sqrt{2}$

$= 22y\sqrt{2}$

13. Solution:

$x\sqrt{2} + 2x\sqrt{2}$

$= (x + 2x)\sqrt{2}$

$= 3x\sqrt{2}$

American Math Academy

15. Solution:

$$2xy\sqrt{2} - xy\sqrt{18}$$
$$= 2xy\sqrt{2} - xy\sqrt{9 \cdot 2}$$
$$= 2xy\sqrt{2} - 3xy\sqrt{2}$$
$$= (2xy - 3xy)\sqrt{2} = -xy\sqrt{2}$$

16. Solution:

$$5\sqrt{2} - 3\sqrt{2} + 8\sqrt{2} - \sqrt{2}$$
$$= \sqrt{2}(5 - 3 + 8 - 1)$$
$$= \sqrt{2}(9)$$
$$= 9\sqrt{2}$$

17. Solution:

$$\sqrt{48xy} - \sqrt{75xy}$$
$$= \sqrt{16 \cdot 3xy} - \sqrt{25 \cdot 3xy}$$
$$= 4\sqrt{3xy} - 5\sqrt{3xy}$$
$$= -\sqrt{3xy}$$

18. Solution:

$$9\sqrt{x} - 3\sqrt{y} + 5\sqrt{x} - \sqrt{y}$$
$$= 9\sqrt{x} + 5\sqrt{x} - \sqrt{y} - 3\sqrt{y}$$
$$= 14\sqrt{x} - 4\sqrt{y}$$

MULTIPLYING AND DIVIDING RADICAL EXPRESSIONS EXERCISES SOLUTIONS

1. Solution:

$$\sqrt{3}\,(2 + \sqrt{8}\,)$$
$$= \sqrt{3}\,(2 + 2\sqrt{2}\,)$$
$$= 2\sqrt{3} + 2\sqrt{6}$$

2. Solution:

$$\sqrt{5}\,(3 + \sqrt{2}\,)$$
$$= 3\sqrt{5} + \sqrt{10}$$

3. Solution:

$$\sqrt{2}\,(5 - \sqrt{7}\,)$$
$$= 5\sqrt{2} - \sqrt{14}$$

4. Solution:

$$\sqrt{10}\,(4 - \sqrt{10}\,)$$
$$= 4\sqrt{10} - \sqrt{100}$$
$$= 4\sqrt{10} - 10$$

5. Solution:

$$\sqrt{3}\,(\sqrt{6} + \sqrt{6}\,)$$
$$= \sqrt{3}\,(2\sqrt{6}\,)$$
$$= 2\sqrt{18} = 2 \cdot 3\sqrt{2} = 6\sqrt{2}$$

6. Solution:

$$\sqrt{8}\,(\sqrt{7} + \sqrt{7}\,)$$
$$= \sqrt{8}\,(2\sqrt{7}\,)$$
$$= 2\sqrt{2}\,(2\sqrt{7}\,)$$
$$= 4\sqrt{14}$$

7. Solution:

$$(\sqrt{2} + \sqrt{3})(\sqrt{2} - \sqrt{3})$$
$$= \sqrt{4} - \sqrt{6} + \sqrt{6} - \sqrt{9}$$
$$= 2 - 3 = -1$$

8. Solution:

$$(\sqrt{a} - \sqrt{b})(\sqrt{a} + \sqrt{b})$$
$$= \sqrt{a^2} - \sqrt{ab} + \sqrt{ab} - \sqrt{b^2}$$
$$= a - b$$

9. Solution:

$$\sqrt{\frac{18}{x^2}} = \frac{\sqrt{18}}{\sqrt{x^2}} = \frac{3\sqrt{2}}{x}$$

10. Solution:

$$\sqrt{\frac{\overset{4}{24}x^4y^8}{\underset{1}{6}x^3y^2}} = \sqrt{4x^{4-3}y^{8-2}}$$

$$= \sqrt{4x^1y^6}$$

$$= 2y^3\sqrt{x}$$

11. Solution:

$$\frac{\sqrt{12}}{\sqrt{2}} = \sqrt{\frac{12}{2}} = \sqrt{6}$$

12. Solution:

$$\frac{\sqrt{8}}{\sqrt{3}} = \frac{2\sqrt{2}}{\sqrt{3}} \cdot \frac{\sqrt{3}}{\sqrt{3}} = \frac{2\sqrt{6}}{3}$$

13. Solution:

$$\frac{\sqrt{48}}{3\sqrt{2}} = \frac{\sqrt{16 \times 3}}{3\sqrt{2}} = \frac{4\sqrt{3}}{3\sqrt{2}} \cdot \frac{\sqrt{2}}{\sqrt{2}}$$

$$= \frac{4\sqrt{6}}{3 \cdot 2} = \frac{4\sqrt{6}}{6} = \frac{2\sqrt{6}}{3}$$

14. Solution:

$$\frac{\sqrt{72}}{6\sqrt{3}} = \frac{\sqrt{36 \times 2}}{6\sqrt{3}} = \frac{6\sqrt{2}}{6\sqrt{3}} = \frac{\sqrt{2}}{\sqrt{3}}$$

$$= \frac{\sqrt{2}}{\sqrt{3}} \cdot \frac{\sqrt{3}}{\sqrt{3}} = \frac{\sqrt{6}}{3}$$

15. Solution:

$$\frac{\sqrt{2} + \sqrt{3}}{\sqrt{2} - \sqrt{3}} = \left(\frac{\sqrt{2} + \sqrt{3}}{\sqrt{2} - \sqrt{3}}\right)\left(\frac{\sqrt{2} + \sqrt{3}}{\sqrt{2} + \sqrt{3}}\right)$$

$$= \frac{\sqrt{4} + \sqrt{6} + \sqrt{6} + \sqrt{9}}{\sqrt{4} - \sqrt{6} + \sqrt{6} - \sqrt{9}}$$

$$= \frac{2 + 2\sqrt{6} + 3}{2 - 3} = \frac{5 + 2\sqrt{6}}{-1} = -5 - 2\sqrt{6}$$

17. Solution:

$$\frac{\sqrt{x} + \sqrt{y}}{\sqrt{x} - \sqrt{y}} = \left(\frac{\sqrt{x} + \sqrt{y}}{\sqrt{x} - \sqrt{y}}\right)\left(\frac{\sqrt{x} + \sqrt{y}}{\sqrt{x} + \sqrt{y}}\right)$$

$$= \frac{\sqrt{x^2} + \sqrt{xy} + \sqrt{xy} + \sqrt{y^2}}{\sqrt{x^2} - \sqrt{xy} + \sqrt{xy} - \sqrt{y^2}}$$

$$= \frac{x + 2\sqrt{xy} + y}{x - y}$$

16. Solution:

$$\left(\frac{\sqrt{5} - \sqrt{7}}{\sqrt{5} + \sqrt{7}}\right) \cdot \left(\frac{\sqrt{5} - \sqrt{7}}{\sqrt{5} - \sqrt{7}}\right)$$

$$= \frac{\sqrt{25} - \sqrt{35} - \sqrt{35} + \sqrt{49}}{\sqrt{25} - \sqrt{35} + \sqrt{35} - \sqrt{49}}$$

$$= \frac{5 - 2\sqrt{35} + 7}{5 - 7} = \frac{12 - 2\sqrt{35}}{-2}$$

$$= -6 + \sqrt{35}$$

18. Solution:

$$\left(\frac{\sqrt{6} + \sqrt{7}}{2\sqrt{6} - \sqrt{7}}\right)\left(\frac{2\sqrt{6} + \sqrt{7}}{2\sqrt{6} + \sqrt{7}}\right)$$

$$= \frac{2\sqrt{36} + \sqrt{42} + 2\sqrt{42} + \sqrt{49}}{2\sqrt{36} + 2\sqrt{42} - 2\sqrt{42} - \sqrt{49}}$$

$$= \frac{12 + 3\sqrt{42} + 7}{24 - 7}$$

$$= \frac{19 + 3\sqrt{42}}{17}$$

RATIONAL EXPONENTS EXERCISES SOLUTIONS

1. Solution:

$6^4 \times 6^5$

$= 6^{4+5}$

$= 6^9$

2. Solution:

$3^{-4} \times 3^{-7} = 3^{-4+(-7)} = 3^{-11} = \dfrac{1}{3^{11}}$

3. Solution:

$(2^3)^{-4} = 2^{-12} = \dfrac{1}{2^{12}}$

4. Solution:

$(a^{-7})^5 = a^{-35} = \dfrac{1}{a^{35}}$

5. Solution:

$(27)^{\frac{-1}{3}} = (3^3)^{\frac{-1}{3}}$

$= 3^{-1} = \dfrac{1}{3}$

6. Solution:

$(81)^{\frac{-1}{5}} = (3^4)^{\frac{-1}{5}} = 3^{-\frac{4}{5}}$

$= \dfrac{1}{3^{\frac{4}{5}}} = \dfrac{1}{\sqrt[5]{3^4}}$

7. Solution:

$\left(\dfrac{1}{27}\right)^{\frac{-1}{3}} = (27^{-1})^{\frac{-1}{3}}$

$= (27)^{\frac{1}{3}} = (3^3)^{\frac{1}{3}}$

$= 3^1$

8. Solution:

$\left(\dfrac{1}{4}\right)^{\frac{-1}{2}} = (4^{-1})^{\frac{-1}{2}}$

$= (4)^{\frac{1}{2}} = (2^2)^{\frac{1}{2}}$

$= 2^1$

9. Solution:

$(4)^{\frac{-1}{3}} = (2^2)^{\frac{-1}{3}} = 2^{-\frac{2}{3}}$

$= \dfrac{1}{2^{\frac{2}{3}}} = \dfrac{1}{\sqrt[3]{2^2}}$

10. Solution:

$(7)^{\frac{-1}{5}} = \dfrac{1}{7^{\frac{1}{5}}} = \dfrac{1}{\sqrt[5]{7^1}}$

11. Solution:

$$(x)^{\frac{-1}{5}} = \frac{1}{x^{\frac{1}{5}}} = \frac{1}{\sqrt[5]{x^1}}$$

13. Solution:

$$(64x^8y^4)^{\frac{-1}{4}}$$

$$= (2^6x^8y^4)^{\frac{-1}{4}}$$

$$= 2^{\frac{-6}{4}} x^{\frac{-8}{4}} y^{\frac{-4}{4}}$$

$$= 2^{\frac{-3}{2}} x^{-4} y^{-1}$$

$$= \frac{1}{2^{\frac{3}{2}} x^4 y^1} = \frac{1}{\sqrt[2]{2^3} x^4 y^1} = \frac{1}{2x^4 y \sqrt{2}}$$

15. Solution:

$$\frac{(3x^3y^6)^{-3}}{y^{-7}} = \frac{3^{-3}x^{-9}y^{-18}}{y^{-7}}$$

$$= \frac{1}{3^3 x^9 y^{18-7}} = \frac{1}{27x^9 y^{11}}$$

17. Solution:

$$\frac{15(x^2y^{-3})^{-6}}{(3^{-1}y^1)^{-3}} = \frac{15(x^{-12}y^{+18})}{3^3 y^{-3}}$$

$$= \frac{15x^{-12}y^{18+3}}{27}$$

$$= \frac{5y^{21}}{9x^{12}}$$

12. Solution:

$$(125)^0 = 1$$

14. Solution:

$$(x^{-10}y^5)^{\frac{-1}{5}}$$

$$= x^{\frac{10}{5}} y^{\frac{-5}{5}}$$

$$= x^2 y^{-1} = \frac{x^2}{y}$$

16. Solution:

$$\frac{(2x^4y^{12}z^{16})^4}{6x^6y^{20}z^{-7}}$$

$$= \frac{2^4 x^{16} y^{48} z^{64}}{6x^6 y^{20} z^{-7}}$$

$$= \frac{\overset{8}{\cancel{16}} x^{16-6} y^{48-20} z^{64+7}}{\underset{3}{\cancel{6}}}$$

$$= \frac{8x^{10} y^{28} z^{71}}{3}$$

18. Solution:

$$\left(\frac{x^3}{\sqrt{x^4}}\right)^{\frac{-1}{2}} = \left(\frac{x^3}{x^2}\right)^{\frac{-1}{2}}$$

$$= (x^1)^{\frac{-1}{2}} = x^{\frac{-1}{2}}$$

$$= \frac{1}{x^{\frac{1}{2}}} = \frac{1}{\sqrt{x}}$$

1) Solution:

$$\boxed{i^2 = -1}$$

$i^{15} + i^{21} + i^{48}$

$= (i^2)^7 i + (i^2)^{10} \cdot i + (i^2)^{24}$

$= (-1)^7 \cdot i + (-1)^{10} \cdot i + (-1)^{24}$

$= \cancel{-i} + \cancel{i} + 1$

$= 1$

2) Solution:

$$\boxed{i^2 = -1}$$

$i^{20} - i^{26}$

$= (i^2)^{10} - (i^2)^{13}$

$= (i^2)^{10} - (i^2)^{13}$

$= (-1)^{10} - (-1)^{13}$

$= 1 - (-1)$

$= 1 + 1 = 2$

3) Solution:

$$\boxed{i^2 = -1}$$

$i^{2020} + i^{2021} + i^{2022}$

$= (i^2)^{1010} + (i^2)^{1010} \cdot i + (i^2)^{1011}$

$= (-1)^{1010} + (-1)^{1010} \cdot i + (-1)^{1011}$

$= \cancel{1} + i - \cancel{1} = i$

4) Solution:

$$\boxed{i^2 = -1}$$

$i^2 - i^{12} + i^{18}$

$= (i^2)^1 - (i^2)^6 + (i^2)^9$

$= (-1)^1 - (-1)^6 + (-1)^9$

$= -1 - 1 - 1$

$= -3$

5) Solution:

$$\dfrac{i^{150}}{i^{19}} \quad \boxed{i^2 = -1}$$

$= \dfrac{(i^2)^{25}}{(i^2)^9 \cdot i} = \dfrac{(-1)^{25}}{(-1)^9 \cdot i} = \dfrac{-1}{-i}$

$= \dfrac{-1\,(i)}{-i\,(i)} = \dfrac{-i}{-i^2} = \dfrac{i}{i^2} = \dfrac{i}{-1}$

$= -i$

6) Solution:

$$\boxed{i^2 = -1}$$

$i^{2020} \times i^{2021}$

$= i^{2020 + 2021} = i^{4041}$

$= (i^2)^{2020} \cdot i = (-1)^{2020} \cdot i$

$= i$

7) Solution:

$$\frac{7}{i} = \frac{7 \cdot i}{i \cdot i} = \frac{7i}{i^2} = \frac{7i}{-1}$$

$$= -7i$$

8) Solution:

$$(i - 2) \cdot (i + 2) = i^2 + 2i - 2i - 4$$

$$= (-1)^1 - 4 = -1 - 4 = -5$$

9) Solution:

$$\frac{3}{2i} - \frac{5}{6i} = \frac{3(3)}{2i(3)} - \frac{5}{6i}$$

$$= \frac{9-5}{6i} = \frac{4}{6i} = \frac{2}{3i}\left(\frac{i}{i}\right)$$

$$= \frac{2i}{3i^2} = \frac{2i}{-3}$$

10) Solution:

$$\boxed{i^2 = -1}$$

$$\sqrt{-625} = \sqrt{625 \times i^2}$$

$$= \sqrt{25^2 \times i^2} = 25i$$

11) Solution:

$$\boxed{i^2 = -1}$$

$$\sqrt{-36}$$

$$= \sqrt{36 \times i^2} = 6i$$

12) Solution:

$$(i - 3) \cdot (i + 5) = i^2 + 5i - 3i - 15$$

$$= (-1) + 2i - 15$$

$$= 2i - 16$$

13) Solution:

$$\left(\frac{3-2i}{5+2i}\right) \cdot \left(\frac{5-2i}{5-2i}\right) = \frac{15 - 6i - 10i + 4i^2}{25 - 10i + 10i - 4i^2}$$

$$= \frac{15 - 16i - 4}{25 - 4(-1)} = \frac{11 - 16i}{29}$$

14) Solution:

$$\left(\frac{1-i}{1+i}\right) \cdot \left(\frac{1-i}{1-i}\right) = \frac{15 - i - i + i^2}{1 - i + i - i^2}$$

$$= \frac{1 - 2i - 1}{1 - i^2} = \frac{-2i}{2} = -i$$

15) Solution:

$$\frac{5-2i}{3} = \frac{(5-2i)\,i}{3i\cdot i}$$

$$= \frac{5i-2i^2}{3i^2}$$

$$= \frac{5i+2}{-3}$$

$$= \frac{-5i}{3} - \frac{2}{3}$$

16) Solution:

$$\left(\frac{-1-2i}{1-3i}\right)\cdot\left(\frac{1+3i}{1+3i}\right) = \frac{-1-3i-2i-6i^2}{1-3i+3i-9i^2}$$

$$= \frac{-1-5i+6}{1+9} = \frac{5-5i}{10} = \frac{1-i}{2}$$

18) Solution:

$$\left(\frac{1+i}{1-i}\right)^{2018} = \left(\frac{(1+i)(1+i)}{(1-i)(1+i)}\right)^{2018}$$

$$= \left(\frac{1+i+i+i^2}{1+i-i-i^2}\right)^{2018}$$

$$= \left(\frac{1+i+i-1}{1+-(-1)}\right)^{2018} = \left(\frac{2i}{1+1}\right)^{2018} = \left(\frac{2i}{2}\right)^{2018}$$

$$= i^{2018} = (i^2)^{1009} = (-1)^{1009}$$

$$= -1$$

17) Solution:

$$\frac{5}{i} = \frac{5(i)}{i(i)} = \frac{5i}{i^2} = \frac{5i}{-1} = -5i$$

SOLVING QUADRATIC EQUATIONS BY FACTORING
EXERCISES SOLUTIONS

1. Solution:

$(x - 7) \cdot (x + 7)$

2. Solution:

$x^2 - 100$

$x \quad x \quad 10 \quad 10$

$(x - 10) \cdot (x + 10)$

3. Solution:

$x^2 - 8x + 15$

$x \quad x \qquad -3 \quad -5$

$(x - 3) \cdot (x - 5)$

4. Solution:

$x^2 - x + 42$

$x \quad x \qquad +6 \quad -7$

$(x - 7) \cdot (x + 6)$

5. Solution:

$6x^2 - 21x - 12$

$6x \quad x \qquad +3 \quad -4$

$(6x + 3) \cdot (x - 4)$

6. Solution:

$x^2 - 2$

$x \quad x \quad \sqrt{2} \quad \sqrt{2}$

$(x - \sqrt{2}) \cdot (x + \sqrt{2})$

8. Solution:

$x^2 + 2xy + y^2$

$x \quad x \qquad y \quad y$

$(x + y) \cdot (x + y)$

7. Solution:

$$\frac{x^2 \cdot x^2}{y^2 \cdot x^2} + \frac{y^2 \cdot y^2}{x^2 \cdot y^2} = \frac{x^4}{x^2 y^2} + \frac{y^4}{x^2 y^2} = \frac{x^4 + y^4}{x^2 y^2}$$

9. Solution:

$$x^2 - 6x - 16$$

x x   +2 −8

$(x + 2) \cdot (x - 8)$

10. Solution:

$$x^2 - 9x + 8$$

x x   −8 −1

$(x - 8) \cdot (x - 1)$

11. Solution:

$$7x^2 + 55x - 72$$

7x x   +8 −9

$(7x + 8) \cdot (x - 9)$

12. Solution:

$$2x^2 - 7x - 15$$

2x x   +3 −5

$(2x + 3) \cdot (x - 5)$

13. Solution:

$$25x^2 + 5x - 2$$

5x 5x   +2 −1

$(5x - 1) \cdot (5x + 2)$

14. Solution:

Not factorable

15. Solution:

$$(x + 3) \cdot (x + 1) = x^2 + x + 3x + 3$$
$$= x^2 + 4x + 3$$
$$k = 4$$

1. Solution:

$$ax^2 + bx + c = 0$$

$a = 1$

$b = 8$

$c = 12$

$$x = \frac{-8 \mp \sqrt{64 - 4(1)(12)}}{2 \cdot 1} = \frac{-8 \mp \sqrt{64 - 48}}{2}$$

$$= \frac{-8 \mp \sqrt{16}}{2}$$

$$x = \frac{-8 \mp 4}{2}, \, x = -6, \, x = -2$$

2. Solution:

$$ax^2 + bx + c = 0$$

$a = 1, \qquad b = 2, \qquad c = 1$

$$x = \frac{-2 \mp \sqrt{2^2 - 4(1)(1)}}{2} = \frac{-2 \mp \sqrt{4 - 4}}{2}$$

$$= \frac{-2}{2} = -1$$

3. Solution:

$$ax^2 + bx + c = 0$$

$a = 4$

$b = 5$

$c = 1$

$$x = \frac{-5 \mp \sqrt{25 - 4(4) \cdot (1)}}{2 \cdot 4} = \frac{-5 \mp \sqrt{25 - 16}}{8}$$

$$= \frac{-5 \sqrt{9}}{8}$$

$$x = \frac{-5 \mp 3}{8}, \, x = -1, \, x = \frac{-1}{4}$$

4. Solution:

$$ax^2 + bx + c = 0$$

$a = 1 \qquad b = -8 \qquad c = 12$

$$x = \frac{8 \mp \sqrt{64 - 4(1)(12)}}{2 \cdot 4} = \frac{8 \mp \sqrt{64 - 48}}{2}$$

$$x = \frac{8 \mp \sqrt{16}}{2} = \frac{8 \mp 4}{2}, \, x = 6, \, x = 2$$

5. Solution:

$$ax^2 + bx + c = 0$$

$a = 1$

$b = -1$

$c = -1$

$$x = \frac{1 \mp \sqrt{(-1)^2 - 4(1)(-1)}}{2 \cdot 1} = \frac{1 \mp \sqrt{1 + 4}}{2} = \frac{1 \mp \sqrt{5}}{2}$$

6. Solution:

$$ax^2 + bx + c = 0$$

$a = 1 \qquad b = -4 \qquad c = 3$

$$x = \frac{4 \mp \sqrt{(-4)^2 - 4(1)(3)}}{2 \cdot 1} = \frac{4 \mp \sqrt{16 - 12}}{2}$$

$$x = \frac{4 \mp \sqrt{4}}{2} = \frac{4 \mp 2}{2}, \, x = 3, \, x = 1$$

7. Solution:

$$ax^2 + bx + c = 0$$

$$a = 1 \qquad b = -6 \qquad c = 1$$

$$x = \frac{6 \mp \sqrt{(-6)^2 - 4(1)(1)}}{2 \cdot 1} = \frac{6 \mp \sqrt{36 - 4}}{2}$$

$$= \frac{6 \pm \sqrt{32}}{2}$$

$$x = \frac{6 \mp 4\sqrt{2}}{2} = 3 \mp 2\sqrt{2}$$

8. Solution:

$$ax^2 + bx + c = 0$$

$$a = 3, \qquad b = 4, \qquad c = -3$$

$$x = \frac{-4 \mp \sqrt{(-4)^2 - 4(3) \cdot (3)}}{2 \cdot 3} = \frac{-4 \mp \sqrt{16 + 36}}{6}$$

$$x = \frac{-4 \mp \sqrt{52}}{6}$$

$$x = \frac{-4 \mp 2\sqrt{13}}{6} = \frac{-2 \mp \sqrt{13}}{3}$$

9. Solution:

$$(x^2 + 4x) + 3 = 0 \qquad (x + 2)^2 - 4 + 3 = 0$$

$$(x + 2)^2 - 1 = 0 \qquad (x + 2)^2 = 1$$

$$x + 2 = \mp 1$$

$$x = -1$$

$$x = -3$$

10. Solution:

$$(x + 3)^2 - 9 - 16 = 0$$

$$(x + 3)^2 = 25, \quad x + 3 = \mp 5$$

$$x = 2 , \ x = -8$$

11. Solution:

$$(x^2 + 12x) + 8 = (x + 6)^2 - 36 + 8 = 0$$

$$(x + 6)^2 = 28 \qquad x + 6 = \mp\sqrt{28}$$

$$x = -6 \mp 2\sqrt{7}$$

12. Solution:

$$(x - 3)^2 - 9 - 20 = 0$$

$$(x - 3)^2 = 29$$

$$(x - 3) = \mp \sqrt{29}$$

$$x = 3 \mp \sqrt{29}$$

13. Solution:

$(2x^2 + 8x) - 10 = 0,$ $2(x^2 + 4x) - 10 = 0$

$2(x + 2)^2 - 8 - 10 = 0$ $2(x + 2)^2 = 18$

$(x + 2)^2 = \sqrt{9}$

$x + 2 \mp 3$

$x + 2 = 3$ or $x + 2 = -3$

$x = 1$ or $x = -5$

14. Solution:

$3(x^2 + 2x) - 18 = 0,$ $3(x + 1)^2 - 3 - 18 = 0$

$3(x + 1)^2 = 21$ $(x + 1)^2 = 7$

$x + 1 = \mp \sqrt{7}$ $x = -1 \mp \sqrt{7}$

15. Solution:

$5(x^2 + 4x) - 40 = 0$

$5(x + 2)^2 - 20 - 40 = 0$

$5(x + 2)^2 = 60,$ $(x + 2)^2 = 12$

$x + 2 = \mp \sqrt{12}$

$x = -2 \mp 2\sqrt{3}$

16. Solution:

$(x^2 + 4x) + 3 = 0$

$(x + 2)^2 - 4 + 3 = 0,$ $(x + 2)^2 = 1$

$x + 2 = \mp 1$

$x = -2 \mp 1$

$x = -3, -1$

1. $A = P(1+r)^t$

$P = \$360$

$r = 0.06$

$t = 5$ years

$A = P(1 + r)^t$

$A = \$360(1 + 0.06)^5$

$A = \$360(1.06)^5$

$A = \$481.76\ldots$

$A \approx \$482$

2. $A = P(1+r)^t$

$P = \$200$

$r = 0.06$

$t = 8$ years

$A = P(1 + r)^t$

$A = \$200(1 + 0.06)^8$

$A = \$200(1.06)^8$

$A = \$318.76\ldots$

$A \approx \$319$

3. $A = P(1 - r)^t$

$P = \$28,000$

$r = 0.06$

$t = 5$ years

$A = \$28,000(1 - 0.06)^5$

$A = \$28,000(0.94)^5$

$A = \$20,549$

4. Checking Account

$= \$358.60 + \$140.20 - \$178.60$

$= \$320.20$

5. $A = P(1 + r)^t$

$P = \$600$

$r = 0.04$

$t = 12$ years

$A = P(1 + r)^t$

$A = \$600(1+0.04)^{12}$

$A = \$600(1.04)^{12}$

$A = \$960.61\ldots$

$A \approx \$961$

6.

	liters pure water	% water	total liters
80% water	x	.80	.80x
30% water	40	.30	.30(40)=12
60% water	x + 40	.60	.60(x+40)

From the last column, you get the equation

$0.80x + 12 = 0.6(x + 40)$ Solve for x.

$0.80x + 12 = 0.6x + 24$

$0.80x - 0.6x = 24 - 12$

$0.2x = 12$

$2x = 120$

$x = 60$ liters.

American Math Academy

7. Let distance $|AD| = 12x$

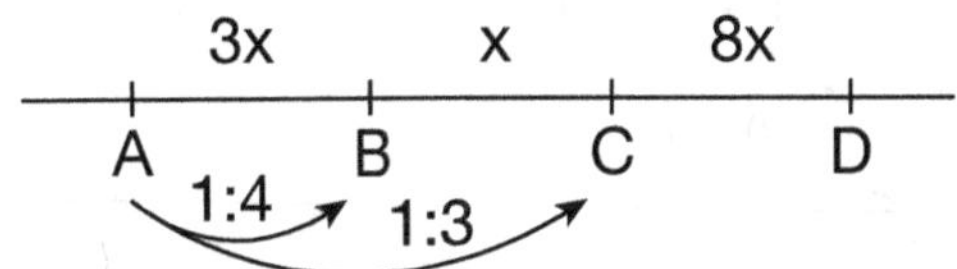

so $|AB| = 3x$

$|AC| = 4x$

$|CD| = 8x$

$|BC| = x$

When the athlete goes from B to C the difference between the times is 10 minutes.

x distance 10 minutes

8x distance 80 minutes

9:00 a.m + 80minutes = 10:20 A.M

8. $\left(\dfrac{120}{60} + \dfrac{80}{60}\right)t = 360$

$\left(\dfrac{200}{60}\right)t = 360$

$\dfrac{10t}{3} = 360$

$\dfrac{t}{3} = 36$, $t = 108$

9. $\dfrac{1}{V} + \dfrac{1}{N} = \dfrac{1}{t}$,

$\dfrac{1}{3} + \dfrac{1}{4} = \dfrac{1}{t}$, then $t = \dfrac{12}{7}$ hours

10. $\dfrac{1}{V} + \dfrac{1}{N} = \dfrac{1}{t}$

$\dfrac{1}{1.5} + \dfrac{1}{3} = \dfrac{1}{t}$, then $t = \dfrac{3}{3} = 1$ hours

1) $a_n = a_1 + (n-1)d \qquad a_{25} = 4 + (25-1)4$

$a_{25} = 4 + (24)4$

$a_{25} = 100$

3) $a_n = a_1 + (n-1)d$

$a_{48} = a_1 + (48-1)12$

$120 = a_1 + (48-1)12$

$120 = a_1 + (47)12$

$120 = a_1 + 564$

$120 - 564 = a_1$, Then $a_1, = -444$

5) For a_{10}

$a_n = a_1 + (n-1)d, \qquad a_{10} = a_1 + (10-1)d,$

$a_{10} = a_1 + (10-1)d$

$48 = a_1 + (10-1)d$

$48 - a_1 = (9)d$

$\dfrac{48 - a_1}{9} = d$

For a_{25}

$a_n = a_1 + (n-1)d, \qquad a_{25} = a_1 + (25-1)d,$

$140 = a_1 + (24)d$

$140 - a_1 = (24)d$

$\dfrac{140 - a_1}{24} = d$

From a_{10} and a_{25}

$\dfrac{48 - a_1}{9} = \dfrac{140 - a_1}{24}$

$9(140 - a_1) = 24(48 - a_1)$

$1260 - 9a_1 = 1152 - 24a_1$

$15a_1 = -108$, then $a_1 = -\dfrac{108}{15} = -\dfrac{36}{5}$

7) $a_n = a_1 (r)^{n-1}$

$a_n = 80(2)^{\frac{1}{2}-1} = 80 \cdot 2^{-\frac{1}{2}} = \dfrac{80}{2^{\frac{1}{2}}} = \dfrac{80}{\sqrt{2}}$

$= \dfrac{80\sqrt{2}}{2} = 40\sqrt{2}$

---

2) $a_n = a_1 + (n-1)d$

$a_{30} = -3 + (30-1)(-7)$

$a_{25} = -3 + (29)(-7)$

$a_{25} = -206$

4) $a_n = a_1 + (n-1)d$

$a_{18} = a_1 + (18-1)8$

$96 = a_1 + (17)8$

$96 = a_1 + (17)8$

$96 = a_1 + 136$, Then $a_1 = -40$

6) $a_n = a_1(r)^{n-1}$

$a_n = 24(8)^{n-1}$

8) $a_n = a_1(r)^{n-1}$

$a_5 = 95(2)^{5-1}$

$a_5 = 95(2)^4$

$a_5 = 95(16)$

$a_5 = 1520$

9) $a_n = a_1 (r)^{n-1}$

$a_7 = 6(2)^{7-1}$

$a_7 = 6(2)^{7-1}$

$a_7 = 6(2)^6$

$a_7 = 6(64)$

$a_7 = 384$

1) Solution:

$$S_n = \frac{n}{2}(a_1 + a_n)$$

$$S_{40} = \frac{40}{2}(5 + 60)$$

$$S_{40} = 1{,}300$$

3) Solution:

$$a_n = a_1 + (n-1)d,$$
$$a_{30} = 10 + (30-1)5,$$
$$a_{30} = 10 + 29 \cdot 5,$$
$$a_{30} = 10 + 145,$$
$$a_{30} = 155$$

$$S_n = \frac{n}{2}(a_1 + a_n)$$

$$S_{30} = \frac{30}{2}(10 + 155)$$

$$S_{30} = 2{,}475$$

5) Solution:

$$r = \frac{10}{5} = 2, \ n = 5$$

$$S_5 = \frac{5(2^5 - 1)}{5 - 1}$$

$$S_5 = \frac{5 \times 31}{4} = \frac{155}{4}$$

7) Solution:

Since $a_n = 2^n - 1$, then 4th term

$$a_4 = 2^4 - 1$$
$$a_4 = 16 - 1$$
$$a_4 = 15$$

9) Solution:

The common difference $x = \frac{1}{2} - 1 = -\frac{1}{2}$

2) Solution:

$$a_n = a_1 + (n-1)d,$$
$$a_{35} = 10 + (35-1)5,$$
$$a_{30} = 7 + 29 \cdot 7,$$
$$a_{30} = 7 + 203,$$
$$a_{30} = 210$$

$$S_n = \frac{n}{2}(a_1 + a_n)$$

$$S_{30} = \frac{30}{2}(7 + 210)$$

$$S_{30} = 3{,}255$$

4) Solution:

$a_1 = 10$, $r = 2$, and $n = 4$

$$S_n = \frac{n(r^n - 1)}{2 - 1}$$

$$S_5 = \frac{4(2^4 - 1)}{2 - 1} = \frac{4(15)}{1} = 60$$

6) Solution:

$$S_n = \frac{a_1(1 - r^n)}{1 - r}$$

$$50 = \frac{a_1(1 - 4^3)}{1 - 4}$$

$$50 = \frac{a_1(1 - 64)}{-3}, \ a_1 = \frac{50}{21}$$

8) Solution:

Arithmetic sequence: 36, 38, 40, ... 74

$$d = 2$$
$$n = 35$$
$$S_n \ \frac{35(36 + 74)}{2}$$
$$= \frac{35 \cdot 110}{2} = \frac{3850}{2} = 1{,}925$$

1) $x^2 + y^2 + 2x - 6y = 20$

$x^2 + 2x + y^2 - 6y = 20$

$(x + 1)^2 - 1 + (y - 3)^2 - 9 = 20$

$(x + 1)^2 + (y - 3)^2 = 20 + 10$

$(x + 1)^2 + (y - 3)^2 = 30$

$(x - h)^2 + (y - k)^2 = r^2$

$r^2 = 30$

$r = \sqrt{30}$

3) $(x - h)^2 + (y - k)^2 = r^2$

$(x - 3)^2 + (y - 6)^2 = (2\sqrt{5})^2$

$(x - 3)^2 + (y - 6)^2 = 20$

5) $\dfrac{(x)^2}{36} + \dfrac{(y)^2}{16} = 1$

We know the equation $\dfrac{(x - h)^2}{a^2} + \dfrac{(y - k)^2}{b^2} = 1$,

the center is (h, k) = (0, 0) and $a^2 = 36$,

a = 6 and $b^2 = 16$, b = 4.

Vertices: (h + a, k) and (h − a, k)

Vertices: (0 + 6, 0) and (0 − 6, 0)

Vertices: (6, 0) and (−6, 0)

2) $x^2 + 10x + y^2 + 4y - 18 = 0$

$(x + 5)^2 - 25 + (y + 2)^2 - 4 = 18$

$(x + 5)^2 + (y + 2)^2 - 29 = 18$

$(x + 5)^2 + (y + 2)^2 = 47$

$(x - h)^2 + (y - k)^2 = r^2$

$r^2 = 47$

$r = \sqrt{47}$

4) $\dfrac{(x)^2}{25} + \dfrac{(y)^2}{9} = 1$

We know the equation $\dfrac{(x - h)^2}{a^2} + \dfrac{(y - k)^2}{b^2} = 1$,

the center is (h, k) = (0, 0) and $a^2 = 25$,

a = 5 and $b^2 = 9$, b = 3.

Distance: $c = \sqrt{a^2 - b^2}$,

$c = \sqrt{5^2 - 3^2} = \sqrt{25 - 9} = \sqrt{16} = 4$

The coordinates of the foci are (h − c, k) and (h + c, k) then (0 − 4, 0) and (0 + 4, 0).

Foci: (−4, 0) and (4, 0).

6) $\dfrac{(x - 2)^2}{25} + \dfrac{(y + 3)^2}{4} = 1$

We know the equation $\dfrac{(x - h)^2}{a^2} + \dfrac{(y - k)^2}{b^2} = 1$,

the center is (h, k) = (0, 0) and $a^2 = 25$,

a = 5 and $b^2 = 4$, b = 2.

Distance: $c = \sqrt{a^2 - b^2}$,

$c = \sqrt{5^2 - 2^2} = \sqrt{25 - 4} = \sqrt{21}$

The coordinates of the foci are (h − c, k) and (h + c, k) then (0 − $\sqrt{21}$, 0) and (0 + $\sqrt{21}$, 0).

Foci: (−$\sqrt{21}$, 0) and ($\sqrt{21}$, 0).

7) $x^2 + 8x + y^2 - 4y + 36 = 0$

$(x + 4)^2 - 16 + (y - 2)^2 - 4 + 36 = 0$

$(x + 4)^2 + (y - 2)^2 - 20 + 36 = 0$

$(x + 4)^2 + (y - 2)^2 + 16 = 0$

$(x + 4)^2 + (y - 2)^2 = -16$

Vertex point $(h, k) = (-4, 2)$

8) $y = x^2 + 6$

$y = (x + 3)^2 - 9 + 6$

$y = (x + 3)^2 - 3$

Since $y = a(x - h)^2 + k$

$h, k = (-3, -3)$

9) When it's open up or down;

$(x - h)^2 = 4p(y - k)$ Where $p \neq 0$ with Vertex point (h, k)

$(x - 3)^2 = 4p(y - 5)$, $h = 3$, and $k = 5$

Vertex point $(h, k) = (3, 5)$

10) When its open right or left;

$(y + k)^2 = 4p(x - h)$ Where $p \neq 0$ with Vertex point (h, k),

Directrix $x = h - p$

$(y + 2)^2 = 4p(x + 7)$, $h = -7$, and $k = -2$

Directrix $x = -7 - 4 = -11$

11) If vertex $= (0,0)$ and focus $= (1,4)$. Since the vertex and focus are $4 - 0 = 4$ units, then $p = 4$

$(x - h)^2 = 4p(y - k)$

$(x - 0)^2 = (4 \cdot 4)(y - 0)$

$x^2 = 16y$

1) Solution:

$\log_5 32 = x$, then $32 = x^5$, $x = 2$

3) Solution:

$\log_3 40 - \log_3 5$

$= \log_3 \dfrac{40}{5} = \log_3 8 = \log_3 2^3$

$= \log_3 2$

5) Solution:

$\log_2 x - 2 + \log_2 x + 3 = \log_2 6$

$\log_2 (x - 2) \cdot (x + 3) = \log_2 6$

$(x - 2) \cdot (x + 3) = 6$,

$x^2 + x - 12 = 0$, $(x-3) \cdot (x + 4) = 0$

$x = 3$ or $x = -4$, since -4 cannot be solution, the x can be only 3.

7) Solution:

$\dfrac{1}{\log_4 8} + \dfrac{1}{\log_4 8} = \dfrac{2}{\log_4 8} = \dfrac{2}{\log_{(2^2)} 2^3}$

$= \dfrac{2}{\frac{3}{2}\log_2 2} = \dfrac{2}{\frac{3}{2}} = \dfrac{4}{3}$

9) Solution:

If $\log_{25} x = \dfrac{3}{2}$, then

$x = 25^{\frac{3}{2}}$, $x = 5^3 = 125$

2) Solution:

$\log_{16}(x) = \dfrac{3}{4}$

$16^{\frac{3}{4}} = x$,

$(2^4)^{\frac{3}{4}} = x$

$2^3 = x$, then $x = 8$

4) Solution:

$\log_{10} 40 - \log_{10} x + 1 = 1$

$\log_{10} \dfrac{40}{x+1} = 1$,

$\dfrac{40}{x+1} = 10^1$, $10x + 10 = 40$, then $x = 3$

6) Solution:

$\log_2 3x - 1 = 5$

$3x - 1 = 2^5$, $3x - 1 = 32$

$3x = 33$

$x = 11$

8) Solution:

$\dfrac{3\log_5 27}{4\log_5 3} = \dfrac{3\log_5 3^3}{4\log_5 3}$

$= \dfrac{9\log_5 3}{4\log_5 3} = \dfrac{9}{4}$

1) Solution:

If $3\begin{bmatrix} x \\ y \end{bmatrix} = 4\begin{bmatrix} 6 \\ 9 \end{bmatrix}$ Divide both side by 3;

$\begin{bmatrix} x \\ y \end{bmatrix} = \dfrac{4}{3}\begin{bmatrix} 6 \\ 9 \end{bmatrix}$, multiply the scalar by each of component.

$\begin{bmatrix} x \\ y \end{bmatrix} = \begin{bmatrix} \frac{4}{3}\cdot 6 \\ \frac{4}{3}\cdot 9 \end{bmatrix}$

$\begin{bmatrix} x \\ y \end{bmatrix} = \begin{bmatrix} 8 \\ 12 \end{bmatrix}$

$x = 8$ and $y = 12$

$x + y = 8 + 12 = 20$

2) Solution:

If $x - y = \begin{bmatrix} 1 & 3 \\ 2 & 4 \end{bmatrix}$, Then $x = y + \begin{bmatrix} 1 & 3 \\ 2 & 4 \end{bmatrix}$

$x = \begin{bmatrix} 3 & 6 \\ 4 & 7 \end{bmatrix} + \begin{bmatrix} 1 & 3 \\ 2 & 4 \end{bmatrix}$, $x = \begin{bmatrix} 4 & 9 \\ 6 & 11 \end{bmatrix}$

4) Solution:

If $\begin{bmatrix} x & y \\ 2 & 4 \end{bmatrix} = 10 = 10$, then $4x - 2y = 10$,

$4x = 2y + 10$

$x = \dfrac{2y}{4} + \dfrac{10}{4} = \dfrac{y}{2} + \dfrac{5}{2}$,

3) Solution:

$A = \begin{bmatrix} a & b \\ c & d \end{bmatrix}$ and $B = \begin{bmatrix} e & f \\ g & h \end{bmatrix}$, Then

$A + B = \begin{bmatrix} a & b \\ c & d \end{bmatrix} + \begin{bmatrix} e & f \\ g & h \end{bmatrix} = \begin{bmatrix} a+e & b+f \\ c+g & d+h \end{bmatrix}$

5) Solution:

$\begin{bmatrix} 1 & 3 \\ -2 & 4 \end{bmatrix}\begin{bmatrix} x \\ y \end{bmatrix} = \begin{bmatrix} 1\cdot x + 3\cdot y \\ -2\cdot x + 4y \end{bmatrix}$

$\begin{bmatrix} 1\cdot x + 3\cdot y \\ -2\cdot x + 4y \end{bmatrix} = \begin{bmatrix} 1 \\ 8 \end{bmatrix}$

$\begin{cases} x + 3y = 1 \\ -2x + 4y = 8 \end{cases}$ Then, $x = -2$ and $y = 1$

$= x + y = -1$

6) Solution:

$x \cdot y = \begin{bmatrix} 1 & 3 \\ -2 & 4 \end{bmatrix}\cdot\begin{bmatrix} 1 & 3 \\ -2 & 4 \end{bmatrix}$

$= \begin{bmatrix} 1\cdot 1 + 3\cdot(-2) & 1\cdot 3 + 3\cdot 4 \\ -2\cdot 1 + 4\cdot(-2) & (-2)\cdot 3 + 4\cdot 4 \end{bmatrix}$

$= \begin{bmatrix} 1-6 & 3+12 \\ -2\cdot -8 & -6+16 \end{bmatrix}$

$= \begin{bmatrix} -5 & 15 \\ -10 & -10 \end{bmatrix}$

American Math Academy

FINAL TEST SOLUTION

1. $\dfrac{\sqrt{5}}{\sqrt{3}} - \dfrac{\sqrt{3}}{\sqrt{5}} = \dfrac{\sqrt{5} \cdot \sqrt{5}}{\sqrt{3} \cdot \sqrt{5}} - \dfrac{\sqrt{3} \cdot \sqrt{3}}{\sqrt{5} \cdot \sqrt{3}} = \dfrac{\sqrt{25}}{\sqrt{15}} - \dfrac{\sqrt{9}}{\sqrt{15}}$

$= \dfrac{5}{\sqrt{15}} - \dfrac{3}{\sqrt{15}} = \dfrac{2}{\sqrt{15}} = \dfrac{2 \cdot \sqrt{15}}{\sqrt{15} \cdot \sqrt{15}} = \dfrac{2\sqrt{15}}{15}$

Correct Answer : B

2. $\dfrac{\sqrt{x}}{2 + \sqrt{x}} = \dfrac{\sqrt{x} \cdot (2 - \sqrt{x})}{(2 + \sqrt{x})(2 - \sqrt{x})} = \dfrac{2\sqrt{x} - x}{4 - x}$

Correct Answer : A

3. $\dfrac{x - 2}{y - 3} = \dfrac{2}{3}$ (cross multiply)

$2(y - 3) = 3(x - 2)$

$2y - 6 = 3x - 6$

$2y = 3x$

$y = \dfrac{3}{2}x$

Correct Answer : D

4. Since the interior of triangle angles are always 180°

The ratio of interior of triangle angles are need to be 180°.

$8x + 6x + 4x = 180°.$

$18x = 180°.$

$x = 10°.$

Large angle = $8x = 8 \cdot 10° = 80°.$

Correct Answer : D

5. $y = \dfrac{k}{x}$ (Inverse variation)

$k = x \cdot y$, then $k = 10 \cdot 18$, $k = 180$.

When x = 36, then yx = k

$y \cdot 36 = 180$

$y = 5$

Correct Answer : C

6. Total students: 20

Students received A: 30% $= \dfrac{30 \cdot 20}{100} = 6$ students

Correct Answer : C

7. $0.015 = \dfrac{15}{1000} = 1.5\%$

Correct Answer : B

8. First find the reciprocal of the fraction then multiply the number by the reciprocal of the fraction.

$\dfrac{5 \cdot x}{2} = \dfrac{1}{8} \cdot \dfrac{2}{3}$

$\dfrac{5x}{2} = \dfrac{2}{24}$ (Cross multiply)

$5x \cdot 24 = 2 \cdot 2$

$120x = 4$

$x = \dfrac{4}{120} = \dfrac{1}{30}$

Correct Answer : B

9. let x is the number of game the team lose.

x = 11 games lose.

Percent of game lose:

$$\frac{11}{55} = \frac{11 \div 11}{55 \div 11} = \frac{1}{5} = \frac{1 \cdot 20}{5 \cdot 20} = \frac{20}{100} = 20\%$$

Correct Answer : A

10. If x is a positive number and

$x^2 + x - 12 = 0$, then $(x - 3)(x + 4) = 0$

$x - 3 = 0$, or $x + 4 = 0$

$x = 3$ or $x = -4$

Correct Answer : C

11. $\frac{x}{4} - 6 < 7$ or $\frac{x}{4} - 6 < -7$

$\frac{x}{4} < 7 + 6$ or $\frac{x}{4} > -7 + 6$

$\frac{x}{4} < 13$ or $\frac{x}{4} > -1$

$x < 4 \cdot 13$ or $x > 4(-1)$

$x < 52$ or $x > -4$

$x < 52$ or $x > -4$

$-4 < x < 52$

Correct Answer : D

12. If $3x - 10 \leq 8$,

$3x - 10 + 10 \leq 8 + 10$,

$3x \leq 18$

$3x \leq 18$ (add 1 both side)

$3x + 1 \leq 18 + 1$

$3x + 1 \leq 19$

Largest possible value of 3x+1 is 19.

Correct Answer : D

13. Since x >0,then

$|x| = x$

$|-2x| = 2x$

$|-3x| = 3x$

So, $\dfrac{|x| + |-2x|}{|-3x|} = \dfrac{x + (2x)}{3x} = \dfrac{3x}{3x} = 1$

Correct Answer : C

14. Use prime factorization to find the greatest prime factor of 26 and 55.

If x is the greatest prime factor of 26, then $26 = 2 \cdot 13$ and the greatest prime factor of 26 is 13. (x = 13)

If y is the greatest prime factor of 55, then $55 = 5 \cdot 11$ and the greatest prime factor of 55 is 11. (y = 11)

$x + y = 13 + 11 = 24$

Correct Answer : C

15. Since the population of a City increased from 270 thousand to 360 thousand

Percent of increase:

$$\frac{360 - 270}{270} = \frac{90}{270} = \frac{9}{27} = \frac{1}{3} = 33.\overline{3}\%$$

Correct Answer : C

16. $\tan\alpha = \dfrac{2}{3}$

$(0, 0)$ is satisfied

$y - 0 = \dfrac{2}{3}(x - 0)$

$y = \dfrac{2}{3}x$

Correct Answer : B

17. $f(x) = t(x + 2)(x - 4)(x + 7)$

So, $x + 7$ is a factor of $f(x)$

Correct Answer : C

18. Even numbers are 0,2,4,6,and, 8.

Probability of the selecting even number

$= \dfrac{5}{10} = \dfrac{1}{2}$

Correct Answer : A

19. $\quad 3y - \dfrac{x}{4} = 10$

$3y - 10 = \dfrac{x}{4} \longrightarrow$ multiply both sides by 2.

$2(3y - 10) = \dfrac{x}{4}(2)$

$6y - 20 = \dfrac{x}{2}$

Correct Answer : A

20. $x^{\frac{1}{3}} = 64$, then $x^{\frac{1}{3}} = 4^3 \longrightarrow$ multiply each power by 3 to find x

$$x^{\frac{1(3)}{3}} = 4^{3(3)}$$

$$x = 4^9$$

$$\dfrac{x}{2} = \dfrac{4^9}{2^1} = \dfrac{(2^2)^9}{2^1} = \dfrac{2^{18}}{2^1} = 2^{18-1} = 2^{17}$$

Correct Answer : B

21. If $x^3 \cdot y^2 > 0$ so y^2 is always positive then $x > 0$

If $x^2 \cdot z > 0$ so x^2 is always positive then $z > 0$

If $y^3 \cdot z < 0$ so we knows $z > 0$ then $y < 0$

Correct Answer : D

22. $= \sqrt{27} - \sqrt{81} + \sqrt{243}$

$= 3\sqrt{3} - 9 + 9\sqrt{3}$

$= 12\sqrt{3} - 9$

Correct Answer : D

23. Discount 20%

Orginal Price = 100x

Discount 20% = 20x

Sale Price = Original Price – Discount

$\$45 = 100x - 20x$

$45 = 80x$

$\dfrac{45}{80} = x$

$x = 56.25\$$

Correct Answer : D

24. $y = k$

$20 = k \cdot 5$

$k = 4$

___

$y = kx$

$y = 4 \cdot 3$

$y = 12$

Correct Answer : C

25. $\dfrac{a^2}{b^2} + \dfrac{b^2}{a^2} = 7$, then

$\dfrac{a}{b} + \dfrac{b}{a} = ?$

$\left(\dfrac{a}{b} + \dfrac{b}{a}\right)^2 - 2 = 7$

$\left(\dfrac{a}{b} + \dfrac{b}{a}\right)^2 = 9$

$\dfrac{a}{b} + \dfrac{b}{a} = 3$

Correct Answer : B

26. $4 + (2 + 3) \cdot 5 - 8 + (5 + 3)^0$

$= 4 + 5 \cdot 5 - 8 + 1$

$= 4 + 25 - 8 + 1$

$= 29 - 8 + 1$

$= 21 + 1$

$= 22$

Correct Answer : D

27. $x - 3 = \sqrt{x - 3}$

$(x - 3)^2 = x - 3$

$x^2 - 6x + 9 = x - 3$

$x^2 - 7x + 12 = 0$

$(x - 3) \cdot (x - 4) = 0$

$x = 3$ or $x = 4$

Correct Answer : D

28. Solve $\dfrac{2}{2 - \sqrt{2}}$

$= \dfrac{2(2 + \sqrt{2})}{(2 - \sqrt{2})(2 + \sqrt{2})}$

$= \dfrac{4 + 2\sqrt{2}}{4 - 2} = \dfrac{4 + 2\sqrt{2}}{2} = 2 + \sqrt{2}$

Correct Answer : C

29. $\dfrac{1}{x - 1} + \dfrac{1}{x + 1} = \dfrac{x + 1 + x - 1}{(x - 1)(x + 1)}$

$= \dfrac{2x}{(x - 1)(x + 1)}$

$= \dfrac{2x}{x^2 - 1}$

Correct Answer : D

30. $\dfrac{x^2 - 8x + 15}{x^2 - 9} \div \dfrac{x^2 - 4x - 5}{x^2 + 3x}$

$\dfrac{(x - 5)(x - 3)}{(x - 3)(x + 3)} \quad \dfrac{x(x + 3)}{(x - 5)(x + 1)}$

$= \dfrac{x}{x + 1}$

Correct Answer : B

31. $x + 3ky = 12$

$5x - 12y = 18$

$$m_1 = \frac{-1}{3k} \; , \; m_2 = \frac{+5}{12}$$

if a system has no solution, slopes are equal

$$m_1 = m_2$$
$$\frac{-1}{3k} = \frac{+5}{12} \; , \; 15k = -12$$
$$k = \frac{-12}{15} = \frac{-4}{5}$$

Correct Answer : A

32. $P(x + 2) = 0, \quad x + 2 = 0, \quad x = -2$

$P(-2) = 2(-2)^3 + a(-2)^2 - (-2) + 2$

$P(-2) = -16 + 4a + 2 + 2$

$0 = -16 + 4a + 4$

$0 = -12 + 4a$

$12 = 4a$

$3 = a$

Correct Answer : B

33. $\boxed{i^2 = -1}$

$(i^2)^{1009} \cdot i + (i^2)^{1010} + (i^2)^{1010} \cdot i$

$(-1)^{1009} \cdot i + (-1)^{2010} + (-1)^{1010} \cdot i$

$= \cancel{-i} + 1 + i$

$= 1$

Correct Answer : B

34. $A = P(1 + r)^t$

$P = \$600$

$r = 0.04$

$t = 12$ years

$A = P(1 + r)^t$

$A = \$600(1 + 0.04)^{12}$

$A = \$600(1.04)^{12}$

$A = \$960 \, . \, 61...$

$A = \$961$

Correct Answer : C

35. $P(x - 2) = x^2 + 2x + 1$

To find P $(x + 1)$ plug in $x + 3$ for x in $P(x - 2)$.

$P(x + 3 - 2) = (x + 3)^2 + 2(x + 3) + 1$

$P(x + 1) = x^2 + 6x + 9 + 2x + 6 + 1$

$P(x + 1) = x^2 + 8x + 16$

Correct Answer : A

36. If $A = 3x + 1 = 4y + 2 = 5z + 3$,

then $A + 2 = 3x + 3 = 4y + 4 = 5z + 5$.

$A + 2 = 3(x + 1) = 4(y + 1) = 5(z + 1)$

$A + 2 = $ LCM $(3, 4, 5) = 60$.

$A + 2 = 60, \quad$ then $A = 58$

Correct Answer : A

37. If d represents the price of the book, then $d - 0.25\,d = 0.75d$ (the price of book after the discount)

If a 4% tax is added, the final price $0.75d + 0.04(0.75\,d) = 0.78d$

Correct Answer : B

38. If $\bigcirc = \triangle$, then

$$x^2 - 2x + 1 = x^2 - 1$$
$$-2x + 1 = -1$$
$$-2x = -2$$
$$x = 1$$

Correct Answer : A

39. $\tan\alpha = \dfrac{2}{3}$

$(0,0)$ is satisfied;

$$y - 0 = \dfrac{2}{3}(x - 0)$$
$$y = \dfrac{2}{3}x$$

Correct Answer : B

40.
$$\dfrac{1}{a} = \dfrac{1}{b} + \dfrac{1}{c}$$
$$\dfrac{1}{a} = \dfrac{c+b}{bc} \text{(cross multiply)}$$
$$ac + ab = bc$$
$$ac = bc - ab$$
$$ac = b(c - a)$$
$$\dfrac{ac}{c-a} = b$$

Correct Answer : A

41.
$$\dfrac{x^2}{y^2} = \dfrac{a^2}{b^2} = \dfrac{4}{9} \text{(Cross multiply)}$$
$$9x^2 = 4y^2 \longrightarrow y^2 = \dfrac{9x^2}{4}$$
$$9a^2 = 4b^2 \longrightarrow b^2 = \dfrac{9}{4}a^2$$

if $y^2 - b^2 = 27$

$$\dfrac{9x^2}{4} - \dfrac{9}{4}a^2 = 27$$
$$\dfrac{9x^2 - 9a^2}{4} = 27$$
$$\dfrac{9(x^2 - a^2)}{4} = 27$$
$$x^2 - a^2 = \dfrac{27 \cdot 4}{9} = 12$$

Correct Answer : A

42. $a = 1 - 5i$ and $b = 1 + 5i$, then

$$a \cdot b = (1 - 5i) \cdot (1 + 5i)$$
$$= 1 - 25i^2$$
$$= 1 + 25$$
$$= 26$$

Correct Answer : D

43. Original price = $100x$.

20% discount from original price

$$= \dfrac{20 \cdot 100x}{100} = 20x.$$

Sales price $= 100x - 20x = \$60$

$$80x = \$60$$
$$x = \dfrac{60}{80} = \dfrac{6}{8} = \dfrac{3}{4}.$$

Original price $= 100x = 100 \cdot \dfrac{3}{4} = \dfrac{300}{4} = 75.$

Correct Answer : D

FINAL TEST SOLUTION

44. $(\cos x + \sin x)^2 = (\sqrt{5} + 1)^2$

$\rightarrow \cos^2 x + 2\cos x \cdot \sin x + \sin^2 x = 5 + 2\sqrt{5} + 1$

$\rightarrow \cos^2 x + \sin^2 x = 1 \Rightarrow$

$1 + 2\cos x \cdot \sin x = 6 + 2\sqrt{5}$, then

$\sin x \cdot \cos x = \dfrac{5 + 2\sqrt{5}}{2}$

Correct Answer : B

45. $A - B = \begin{bmatrix} a & b \\ c & d \end{bmatrix} - \begin{bmatrix} e & f \\ g & h \end{bmatrix} = \begin{bmatrix} a-e & b-f \\ c-g & d-h \end{bmatrix}$

Correct Answer : D

46. If $x^3 = y \Rightarrow \log_y x^2 = \log_{x^3} x^2 = \dfrac{2}{3}$, $\log_x x = \dfrac{2}{3}$

Correct Answer : D

47. $\begin{bmatrix} a & 6 \\ b & 4 \end{bmatrix} = 18 \Rightarrow 4a - 6b = 18$

$4a - 6b = 18$ (divided each term by 2)

$2a - 3b = 9$

$a = \dfrac{9 + 3b}{2}$

Correct Answer : A

48.

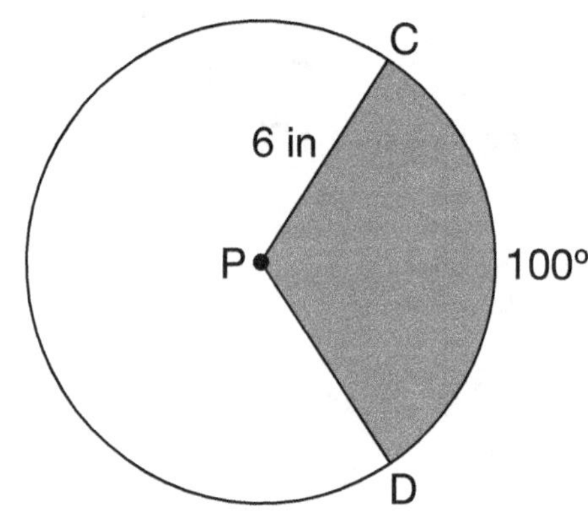

Shaded Area $= \dfrac{\pi r^2 \alpha}{360°}$

$= \dfrac{\pi \cdot 6^2 \cdot 100°}{360°}$

$= \dfrac{36\pi \cdot 100°}{360°}$

$= 10\pi$

Correct Answer : B

AMERICAN MATH
ACADEMY

49. $\dfrac{a^{2m}}{a^{10}} = a^6$

$a^{2m-10} = a^6$

$2m - 10 = 6$, $2m = 16$

$m = 8$

$a^{3n} = a^{30}$, $3n = 30$ $n = 10$

$m \cdot n = 8 \cdot 10 = 80$

Correct Answer : D

50. $\dfrac{1}{3}(6x - 9) + (x - 12) = ax + x + b$

$2x - 3 + x - 12 = ax + x + b$

$3x - 15 = x(a + 1) + b$

$b = -15$

$3x = x(a + 1) \Rightarrow a + 1 = 3$, $a = 2$

$a - b = 2 - (-15)$

$= 17$

Correct Answer : D

51. $(x^2y - 2x^2 + 6xy) - (xy^2 - 2x^2 + 4xy)$

$= x^2y - 2x^2 + 6xy - xy^2 + 2x^2 - 4xy$

$x^2y - xy^2 + 2xy$

Correct Answer : A

52. $\dfrac{2\cdot 3}{2\cdot 4}x - \dfrac{1}{8}x = \dfrac{1}{12} + \dfrac{2}{3}$

$\dfrac{6x - x}{8} = \dfrac{9}{12}$

$\dfrac{5x}{8} = \dfrac{3}{4}$ $x = \dfrac{6}{5}$

Correct Answer : D

53. $i^2 = -1$

$\dfrac{3+i}{2-i}$

$= \left(\dfrac{3+i}{2-i}\right)\left(\dfrac{2+i}{2+i}\right)$

$= \dfrac{6+3i+2i+i^2}{4-i^2}$

$= \dfrac{6+5i-1}{4+1}$

$= \dfrac{5+5i}{5} = 1+i$

Correct Answer : D